Medicine and Biomedical Sciences in Modern History

Series Editors
Carsten Timmermann, University of Manchester, Manchester, UK
Michael Worboys, University of Manchester, Manchester, UK

The aim of this series is to illuminate the development and impact of medicine and the biomedical sciences in the modern era. The series was founded by the late Professor John Pickstone, and its ambitions reflect his commitment to the integrated study of medicine, science and technology in their contexts. He repeatedly commented that it was a pity that the foundation discipline of the field, for which he popularized the acronym 'HSTM' (History of Science, Technology and Medicine) had been the history of science rather than the history of medicine. His point was that historians of science had too often focused just on scientific ideas and institutions, while historians of medicine always had to consider the understanding, management and meanings of diseases in their socio-economic, cultural, technological and political contexts. In the event, most of the books in the series dealt with medicine and the biomedical sciences, and the changed series title reflects this. However, as the new editors we share Professor Pickstone's enthusiasm for the integrated study of medicine, science and technology, encouraging studies on biomedical science, translational medicine, clinical practice, disease histories, medical technologies, medical specialisms and health policies.

The books in this series will present medicine and biomedical science as crucial features of modern culture, analysing their economic, social and political aspects, while not neglecting their expert content and context. Our authors investigate the uses and consequences of technical knowledge, and how it shaped, and was shaped by, particular economic, social and political structures. In re-launching the Series, we hope to build on its strengths but extend its geographical range beyond Western Europe and North America.

is intended to supply analysis and stimulate debate. All books are based on searching historical study of topics which are important, not least because they cut across conventional academic boundaries. They should appeal not just to historians, nor just to medical practitioners, scientists and engineers, but to all who are interested in the place of medicine and biomedical sciences in modern history.***Medicine and Biomedical Sciences in Modern History***

Natasha Szuhan

The Family Planning Association and Contraceptive Science and Technology in Mid-Twentieth-Century Britain

palgrave
macmillan

Natasha Szuhan
School of Sociology
Australian National University
Canberra, Australian Capital
Territory, Australia

ISSN 2947-9142 ISSN 2947-9150 (electronic)
Medicine and Biomedical Sciences in Modern History
ISBN 978-3-030-81299-7 ISBN 978-3-030-81300-0 (eBook)
https://doi.org/10.1007/978-3-030-81300-0

Cover image: Katrinka Szuhan

This Palgrave Macmillan imprint is published by the registered company Springer Nature Switzerland AG
The registered company address is: Gewerbestrasse 11, 6330 Cham, Switzerland

For Mum, Dad, Katrinka and Michael.

Acknowledgements

This work was supported by an Australian Government Research Training Scholarship, and the Wellcome Trust, University of Strathclyde and Shanghai University through a Strathclyde/Shanghai Early Career Medical Humanities Fellowship.

It would not have been possible without the generous guidance and criticism of Joy Damousi, Patricia Grimshaw, Robert Reynolds, Phillipa Levine, Barbara Brookes, Jim Mills and all my fabulous colleagues at the University of Melbourne, Australian National University, University of Strathclyde and Shanghai University.

Contents

Abbreviations

BCIC	Birth Control Investigation Committee
BMA	British Medical Association
BMJ	*British Medical Journal*
BSI	British Standards Institution
CIFC	Council for the Investigation of Fertility Control
FDA	Food and Drug Administration
FPA	Family Planning Association
NBCA	National Birth Control Association
NBCC	National Birth Control Council
NHS	National Health Service
NKWWC	North Kensington Women's Welfare Centre
SPBCC	Society for the Provision of Birth Control Clinics
WWWC	Walworth Women's Welfare Centre

CHAPTER 1

Introduction

Abstract At the turn of the twentieth century, three social and scientific theories intersected in public, political and moral debates in Britain: Social Darwinism, Malthusianism and Eugenics. Each supported scientific birth control to address decreasing health and living standards and population concerns. To each, there was only one guaranteed remedy to these problems: birth control. In response to these needs, a National Birth Control Association was founded that in collaboration with scientists, physicians and Eugenists pioneered a field of contraceptive science to understand, test and guarantee contraceptive practices and technologies.

The Family Planning Association (FPA), a British-founded, but now global, organisation recently celebrated ninety years providing sex and fertility education, advice and therapies. Its current iteration that focuses heavily on knowledge-provision and advice is vastly different from its original therapeutic services. This is because services must adapt to changing socio-cultural needs, norms and knowledge. Now, sex and contraception are demystified and studied as legitimate medical, scientific, social and cultural phenomena, but the association's role in effecting this change has been little studied and thus under-appreciated. This book aims to remedy that omission and place this achievement in its proper social, scientific, medical, educational, and lobbying and networking contexts.

N. Szuhan, *The Family Planning Association and Contraceptive Science and Technology in Mid-Twentieth-Century Britain*, Medicine and Biomedical Sciences in Modern History, https://doi.org/10.1007/978-3-030-81300-0_1

In 1931 the association's predecessor, the National Birth Control Association (NBCA) was founded. This group intended to positively address the 'general problems of population' via the introduction of medical, pure and applied scientific and educational programs at affiliated birth control clinics. The 'practice of contraception' was then becoming 'increasingly important' to local, national and global interests and pursuits, as eugenic and dysgenic populations, as well as steep birth rate decline, became key foci for governments, doctors, eugenists and philanthropists keen to address social and political issues of demographic, sexual, educational and medico-scientific character.[1]

This is not an exciting history of a great socio-cultural change in the understanding or application of contraceptives in Britain. Histories of the 'progress' and 'success' of a variety of birth control methods, practices and providers were staples of the late twentieth century. This book tells a more nuanced and humble history of the association and its members as leaders in medical contraceptive provision, scientific research, application and advocacy, the creation and delivery of education, advice and therapy, and networking and lobbying for the acceptance and oversight of contraception by officials, both medical and governmental. It attempts to complicate and extend understandings of the working and function of the contraceptive lobby and its interactions with government and approved regulators, the medical and scientific professions, contraceptive knowledge and product manufacturers and providers in the interwar and post-war periods and to place the organisation into some competitive and cooperative transnational context through its dealings with the American contraceptive research lobby.

It takes as its focus the central organisation and its dedication to the ideal of standardisation. In this context, standardisation relates to the process of ordering and regulating contraceptive products and practices, the management and function of clinics, and sex and birth control knowledge and education, and to taking this regulation as a top-down association-wide aspiration to be extensively disseminated by and from the centralised NBCA/FPA medical and executive management. This particular aspiration was only made possible through the alignment of contraception, medicine and science that began in the mid-1920s and was applied and extended by the association, its contemporary the Birth Control Investigation Committee (BCIC) that took research and later advocacy as its focus, and other national and international groups;

including the Society for the Provision of Birth Control Clinics, Malthusian League, Marie Stopes' Society for Constructive Birth Control and Racial Progress, the Eugenics Society, the Birth Control International Medical Group for the Investigation of Birth Control, Planned Parenthood Federation of America, the American National Committee on Maternal Health and Bureau of Social Hygiene, and Committee for Research in Problems of Sex. The contemporary appreciation that scientific sex research could be applied to contraception to legitimise and guarantee the practice has been established and touched on by earlier historians of chemistry and biology, but the British focus on such studies has been limited and has thus ignored the vital role the association played in the legitimisation and formalisation of the union between science, medicine and contraception.[2]

But how was this achieved in an environment where contraception was considered to be a non-medical and thus non-scientific selfish social pursuit and practice? This will be investigated by charting the association's navigation through the murky, often sexist worlds of science and medicine to understand how and why the mostly female medical staff of its contraceptive clinics and medical and executive committees took the practical, investigative, educational and propagandist directions they did. Often this was to identify an area of medical or social need, scientifically frame and investigate it, and then ruthlessly exploit it to force it into local, national and professional focus in an attempt to address and treat it.[3] This is essentially the process of medicalisation that feminist and medical historians have increasingly focused on and identified since the last quarter of the twentieth century.[4] This has primarily been framed as a means of developing social understanding and influence for the purpose of effecting social controls, both for general social and health improvements, and sometimes for more nefarious methods of coerced control, a la the broad and all-encompassing nature of Foucault's theory of biopower.[5]

But in the particular instance of the association's efforts to medicalise contraception, the process doesn't fit more traditional concepts of the process that frame it as an institutionalisation as that implies the existence of professional support for that eventuality. As Lara Marks has argued, the profession on the whole, radiated an aura of utter contempt towards a medicalised contraceptive practice until the 1960s when it was forced to oversee and prescribe the pill.[6] Up until then, it was more a matter of individual doctor's conscience and interest, and the older generation of physicians were only incrementally convinced that contraception constituted a legitimate preventive therapy. Therefore, the process of medicalisation that this book describes is one that worked both within the local professional medico-scientific arenas and outside it; that was because the association's early work in this space was deemed illegitimate and so it had to forge a valid niche for its research and findings before any formal medicalisation could be effected. Some earlier historians have discussed the process of medicalisation and the use of science to legitimate this field, but most have either jointly focused on the American and global story or have focused on medico-scientific work that directly begat the pill.[7] This early work is the prime focus of this study. I aim to place the association's labour into an intersecting socio-cultural, medico-scientific and intellectual environment where the dynamics of time, place, law, need, gender, sex and often serendipity coalesced to foster and achieve contraceptive agitation and action.

The organisation's work was always client centred and this ultimately meant it was highly gendered in its approach and function. This reality was recently highlighted by Caroline Rusterholz, although she primarily stressed how female contraceptive scientists and physicians, functioning across Britain and France, adapted their personal and professional experiences to contraceptive education and provision within distinct local, national and transnational collaborative cultures.[8] The association tapped into a need to awaken, inform and excite British women's awareness of their reproductive and health needs, and created and heavily pushed female-centric contraceptives to their captive audience of clients—however broad or limited its actual influence may have been at any given time or locale. This was and has subsequently been framed as a means of providing women agency over their body's reproductive and maternal functions, but there is in this approach a tendency, to which I likely succumb on occasion, to frame this as having a more general success than it may have—given the resource and influence limitations the association laboured under for most of its early history.[9] The patient focus of contraceptive clinics has been an increasing academic emphasis in recent years, but the patient experience in these sites is often more difficult to establish and effectively relay than the goals, aspirations and assessments of contraceptive doctors and providers who are still the mediators for patient experiences and voices through their better kept and accessed records.[10] The association's records, on which I rely due to this very necessity, most certainly relay that the organisation had huge aspirations and a deeply held sense of responsibility to women, society and science that informed its work between 1931 and 1972 when the National Health Service finally undertook to provide contraception through formal health networks and sites. However, its patient voices and personal contraceptive and fertility aims remain conspicuously absent, and likely will remain so until the Wellcome Library is able to release the next round of FPA archives, some of which specifically relate to patient records, experiences and complaints in a few more years.

Testing and Regulating Contraceptive Technologies: Contemporary Historiography

Much attention has been paid to the various methods of contraception that have been trialled and tested throughout history. Indeed many of the methods and technologies that will be the focus of this work were available long before the 1930 origins of the association and had been proved effective over centuries—and millennia—of use and remain staples in the modern contraceptive repertoire. These contraceptives and the history of their development use and regulation can be broken down into three types: mechanical methods, chemical methods which were often used in tandem and socio-sexual methods.

Mechanical methods include condoms and sheaths which work by covering the erect penis to prevent sperm entering the vagina; the various types of caps and the diaphragm each protect the entrance of sperm into the uterus by acting as an internal mechanical barrier; sponges work on the same premise as the cap or diaphragms, they are inserted into the vagina and form a barrier to the entry of sperm into the womb; and intrauterine devices (IUD) that are made from gold, silver, stainless steel or plastic and are inserted into the body or neck of the womb depending on the design and prevent a fertilised egg from attaching and developing therein. This was the first contraceptive technology designed to provide long-term reversible birth control. All except the latter were designed to be used with a spermicidal cream, paste or jelly; these work by being applied into the vagina before intercourse and aim to kill or arrest all sperm before they can enter the woman and potentially fertilise an egg. Each has been the focus of some broad and specific historical studies, but most works to date, that take mechanical technologies as their focus, with few exceptions, have been socio-cultural studies, rather than biographies of science and technology; and few have taken Britain as their focus.[11]

The social history of contraception is undoubtedly more exciting and has more far-reaching implications than a highly specialised and technical account of one technology or another—hence studies of the condom and IUD downplay the mechanics of design, production, fallibility, testing, fit and disposal in favour of a sexier, often more relatable and narratable account of the mis/application, sale, legality, morality and knowledge of sexual, reproductive and contraceptive aids.[12]

The history of mechanical contraceptives, especially their standardisation and regulation, is surprisingly limited. Relatively little is known about the control and efficacy of social and mechanical contraceptive, and, without a keen focus, one would be forgiven for thinking that such work originated with the hormonal oral pill when doctors and scientists and regulatory agencies such as the Food and Drug Administration in the United States intervened to manage the experimental and relatively untested drug as it was marketed and ingested en masse.[13] Junod and Marks in particular have demonstrated the transnational British and American (and arguably global) standardisation of oral contraceptives that formalised contraceptive regulation in both countries, and effectively established the regulatory standards that are assumed and demanded today.[14]

There is little attention paid to the longevity of contraceptive regulation that I will demonstrate was a well-established norm before the pill, this is a by-product of the perceived importance of oral contraceptives to contemporary history and society. Every account of the drug imbues it with some fundamental importance to humanity; ranging from the attribution of its pharmaceutical importance as 'the pill' in a sea of medications available in pill form,[15] to it being dubbed 'one of the seven wonders of the modern world'.[16] Its importance as an example of human achievement has gone so uncontested that according to Elaine Tyler May its bearing on late twentieth-century history cannot be overstated, as its very existence impacted many of the 'most dramatic and contentious issues':

> the quest for reproductive rights; challenges to the authority of medical, pharmaceutical, religious, and political institutions; changing sexual mores and behaviours; reevaluation of foreign policy and foreign aid; and women's emancipation.[17]

This elevation hinders historical research, implicitly rendering all that came before insignificant by comparison. The pill's importance is measured on par with the 'discovery of fire, and development of tool-making, hunting, agriculture, urbanism, scientific medicine, and nuclear energy'.[18] As a direct result, the historiographical origins of the history of contraceptive standardisation begin later than the historical record reveals, the late 1950s, rather than the 1920s when the process actually began in earnest. This is an omission address in this book by focusing on one organisation's dedicated work to achieve this end.

Its first thirty or so years, the NBCA/FPA was dedicated to standardising its products and practices before the advent of oral contraceptives. In addition, from 1955 the International Planned Parenthood Federation's medical committee and testing sub-committee each adapted some of the association's methods and developed their own to assess each available contraceptive technology and determine a universal testing methodology.[19] Without the context of this work, the implementation of contraceptive standards during the 1960s and 1970s occurs without any impetus except hormonal contraception. This is inadequate to explain the introduction of sex education and rubber standards concurrently with those covering pharmaceutical and chemical contraception.

The association's scientific research into contraception was not exclusively focused on standardising and controlling chemical and mechanical technologies, but also on scientific efforts to develop a simple and effective contraceptive that 'the stupidest and therefore most undesirable members of society' could apply with ease. This was significant given the 'clandestine and unregulated' nature of the early twentieth-century British contraceptive trade. This was not the sole focus of the association, it too was a pursuit of transnational and global birth controllers. US efforts to achieve this end have been well studied, but the British, and most particularly, the NBCA/FPA work have been mostly ignored, except in relation to eugenics and population decline.[20]

The British laboratory-based spermicide evaluations between 1929 and 1937 have been the focus of some studies. As I will demonstrate, this work was fundamental to the NBCA/FPA's standardisation activities and resulted in the development of Volpar, 'a powerful spermicide, non-irritating, inexpensive, small, solid … and odourless'.[21] However, earlier accounts frame this work without attending the NBCA despite it being deeply involved; it defined and enforced the parameters of an ideal contraceptive based on patient needs, and demanded clinical and medical experience and testing to guarantee quality, efficacy and safety.

Until the close of the 1930s contraceptive research was primarily the interest of the 'social activist rather than the research scientist or practicing physician'.[22] Contraceptive campaigners' social, political and benevolent motivations tainted the professionalisation of the field as 'applied' contraceptive, rather than 'pure' reproductive, science.[23] The few scientists willing to undertake this research were forced to carefully navigate the two spheres within which they worked: the contraceptive lobby, which funded their work, and wanted them to publicly advocate the practice, and the scientific establishment, that ultimately determined their career trajectory and success. In response, Adele Clarke asserts that reproductive scientists established a 'quid pro quo' with contraceptive advocates, whereby the latter ceased demanding scientific research into mechanical contraceptives on the proviso that researchers would develop scientific contraceptives.[24] Although it has been asserted that in America 'chemical contraceptives never summoned the interests of scientists as the pill was to do', this is not true of Britain, where the NBCA/FPA heavily invested in researching and standardising chemical and mechanical contraceptives, and actively and committedly employed physicians and scientists to undertake that work between 1929 and 1959.[25]

As so often happens in the history of science, technology and medicine when one researcher hits on a brilliant new idea or topic, so do several of his or her colleagues. This same phenomenon has recently occurred in the study of the British history of contraception in the early to mid-twentieth century. In recent years several emerging historians each discovered a fascinating series of sources and omissions in the historical record as they related to contraception in the early twentieth century; that there was a culmination of social, political, production and consumption, educational and scientific, and technological and medical efforts being undertaken by doctors, philanthropists, activists, scientists, manufacturers and lay people that together were standardising contraceptive methods and practices, knowledge and acceptability. So, three recent studies have each considered how contraception came to be scientific and standardised in the clinic, the factory, the curriculum, the marketplace, the laboratory and under the law.

The manufacturing and selling of contraception have been curiously absent from the British history of contraception. It was only recently that the commercial production and distribution of rubber goods and to some degree chemical contraceptives have been attended by historians.[26] The business of birth control that emerged and flourished in the interwar years undoubtedly had significant impacts on local moral, marital, sexual and contraceptive cultures—but until very recently this was taken for granted and not proven. The marketplace of contraceptive products and ideas that has been investigated by Borges and Jones places these developers and industrialists at the frontline of technological and manufacturing innovation and regulation, and demonstrates the collaborative and cooperative, but also suspicious and competitive, nature of the relationships between manufacturers, contraceptive product merchants and providers and birth control lobbyists and propagandists. All of these tensions are borne out in this study of the Family Planning Association and its direct and indirect relationship to industry, oversight, propaganda and practice.

The highly gendered nature of contraceptive medicine and activism has also recently come into focus, as Rusterholz has identified that it was female activist doctors at the vanguard of the contraceptive movement's professionalisation and ultimate success.[27] That females had a particular interest and investment in reproductive and maternal medicine and education and guaranteeing the efficacy, safety and standards of contraception has long been asserted; these aims fit very clearly and logically with contemporary aspirations for female enfranchisement and liberation from the limits of their sexed bodies.[28] But the central role of association-affiliated women doctors and volunteers in the medical research and educational work of groups like the NBCA/FPA, Society for the Provision of Birth Control Clinics (SPBCC), the BCIC and Marie Stopes' society and their responsiveness to identified patient needs has not previously been central to historical studies.[29]

This book will build on and extend all of these studies by foregrounding the Family Planning Association and its clinics as a primary driver and site of the social and medico-scientific changes that were effected during this period. It will chart how the association worked to institute and implement top-down policies and procedures that were to be implemented, trialled and their results fed back to the central management to effect a replicable contraceptive service that could be and was rolled out across the United Kingdom. In building this argument, I focus heavily on the medical and executive committees as these were the association's prime sites of regulatory hypothesis and trial—it was these groups that collaborated to develop and incorporate the medico-scientific standards that the association implemented and widely exported.

This work places the NBCA/FPA at the centre of the British story, to highlight the long-term pioneering role it had in the establishment of standards relating to contraceptive products, provision and education.

Historical and Cultural Context: Competing Theories of Population and Birth Control

Before 1960 no single contraceptive technology could guarantee protection from an undesired pregnancy and so the fallibility of birth control was accepted by couples who knew of available methods and chose to apply them.[30] In the 1920s methods to test, prescribe and guarantee these technologies were hypothesised, developed and applied in the United States, the United Kingdom and its developed colonies. This work was a by-product of the intersection and competition of three social and scientific proposals for ideal public, political and moral organisation, standardisation and control: Social Darwinism, Malthusianism and Eugenics.

Malthusianism, a theory articulated in 1798 by Thomas Malthus taught that nature implemented population 'checks' to ensure human (and animal) populations remained in balance with resources. Malthus identified 'preventive checks' as the socio-sexual methods sentient beings employed to control population, such as delayed marriage, abstinence and preventing 'promiscuous intercourse'; and that 'positive checks' such as war, extreme poverty, epidemic disease and famine would be necessitated if populations surpassed resource 'carrying capacity'.[31] During the nineteenth century, these principles were applied to understand social and population problems: poor laws, medical intervention, degeneracy and optimal population size and health.[32] The neo-Malthusian movement promoted birth control as an ideal preventive check for broad application.[33]

Social Darwinism was a social and political idea that extrapolated Charles Darwin's theory of evolution to human populations. Darwin's philosophy postulated that all living things obeyed one key universal law in the drive to live and multiply: 'let the strongest live and the weakest die'.[34] This idea, that nature's laws predetermined some creatures would fail to flourish, reproduce or survive, was irresistible, and was applied to human societies at the turn of the twentieth century to justify social, economic and health policies that argued against intervention to alleviate 'the poverty of the incompetent, the distress of the careless, the elimination of the lazy' and to argue that the 'pressure of the strong which sweeps aside the weak and reduces so many of them to misery, was a 'necessary result of an enlightened and beneficial general law' of nature.[35] Thus human struggle was an entirely natural outcome of economic competition and success or failure in this environment. In fact, it was Darwin's theory in action. The idea underpinned British advocacy for the 'elimination' of dependents on the state by removing welfare provisions that had increased the population of 'the disabled and the poor' by keeping them just healthy enough to breed more dependents.[36] Subscribers to social Darwinist ideals, and thus anti-welfare movements, proposed to remove the social, financial and moral burden of dependent classes, and supported 'birth control [to] cut public costs and thus taxes'.[37]

British social and population policy and welfare efforts were also influenced by eugenics. This was the most renowned, but also maligned, population control ideology. Eugenic science employed aspects of natural selection to contend that both positive and negative family traits were hereditary.[38] An interest in genetic and social inheritance led Francis Galton to develop and apply 'scientifically sound [welfare and population] policies' and formulate a 'science of improving [human] stock': eugenics.[39] Eugenists argued for two-strand, heredity-based population policies to make qualitative improvements to populations through 'selective breeding'; positive and negative eugenics.[40] The former 'encourage[d] the fittest of all classes' to breed their 'rejuvenating progeny' and the latter discouraged 'the excessive fertility of the labouring poor'.[41] These dual policies continued to be applied to twentieth-century eugenics-based educational programs and informed and shaped the alliance between the eugenics and contraceptive lobbies.

Adherents of each philosophy were seriously concerned about declining national health and living standards, and the rising population of poor and degenerate Britons. All agreed that effective birth control could remedy these problems but worried that the poor, uneducated masses couldn't effectively apply modern contraceptives. By 1920 diverse birth control methods were available, but no single form of contraception could guarantee temporary sterility. Mechanical and chemical methods always had to be paired to ensure some semblance of efficacy by combining a rubber or plastic barrier with a spermicide. Other methods, such as the intra-uterine device that rendered the uterus inhospitable were then being trialled, but their abortifacient qualities rendered the method morally ambiguous, as well as dangerous. Most popular were socio-cultural methods such as abstinence and coitus interruptus or withdrawal, but these gambled on a confluence of motivation, control, precision and luck.[42] The difficulty and uncertainty involved with these methods rendered each capricious and necessitated a scientific intervention to guarantee socially desired contraceptive outcomes.

In response the Malthusian League, a pro-birth control collective was formed in 1877.[43] It promoted family limitation via mechanical and moral contraceptive instruction that it guaranteed as factual and effective.[44] The league initially opted to focus 'its propaganda campaign' widely to awaken social awareness of birth control rather than advocating a particular method or addressing only a specific class, but its effects were most clearly observed amongst the middle class.[45] It promoted and prescribed a variety of contraceptive methods and techniques that it could monitor for efficacy including 'natural' methods, the 'safe' or rhythm method and withdrawal, and 'unnatural' methods including sheaths, the vaginal sponge 'dipped in vinegar' solution and also douching immediately after sex.[46]

Around 1900, birth control, Eugenics and heredity came together in a drive to promote 'scientifically legitimated social technologies'.[47] Eugenists sought to advertise the positive and negative effects of heredity, but found no favour with the Malthusian League that advocated universal fertility limitation rather than differential application depending on fitness.[48] The eugenists thus determined to form a competing group: the Eugenics Education Society in 1907 to prevent social, moral and physical population decay and overpopulation.[49] But birth control fit awkwardly within the eugenic ideology; it was both a great threat to national eugenics, as well as a potential saviour. Population decline amongst the 'fit' was, correctly or otherwise, attributed to increased contraceptive knowledge and application and the attractiveness of small families to achieve higher quality of life.[50] The opposite was true for working and lower classes. This necessitated the society's 'dual [positive and negative eugenics] campaign'; and the more pressing need for negative eugenics obliged the society to focus its efforts on providing 'facilities for birth control'. This eugenic goal persevered into the mid-1920s when the Eugenics Society was formed and allied with local birth controllers, primarily the NBCA, to facilitate and subsidise scientific research and development into contemporary contraceptive safety and efficacy and develop superior methods for mass application.[51]

By 1920, both the League and Education Society were declining, just as birth control was becoming widely known and a nascent contraceptive consumer base was emerging. This was the time manufacturers, providers and advocates began to invest great energy and funds into the development and refinement of contraceptive technologies. The prior century saw barrier-based contraceptives such as sheaths and caps increasingly popularised and made cheaper and more accessible through rubber industry expansion and advances that facilitated the mass production.[52] These were huge advances in contraceptive technology but the convenience and value of birth control remained limited by the need to apply two methods at each intercourse to sufficiently guarantee efficacy—discovering or developing a single-use method was the contraceptive Holy Grail.[53]

Between 1921 and 1925 Britain's first three birth control clinics were opened in poor London suburbs. In March 1921 birth control pioneer Marie Stopes opened the first 'Mothers Clinic' in North London through her Society for Constructive Birth Control and Racial Progress. Within the year, the Malthusian League opened its clinic in Walworth. Finally, the Society for the Provision of Birth Control Clinics opened the North Kensington Women's Welfare Clinic. Each group was sympathetic to Malthusianism, social Darwinism and eugenics, and aimed to provide contraceptive education and products to the poor.[54]

By 1930 the Malthusian league had collapsed, and the SPBCC took over its clinic. That year the society joined with Marie Stopes, and several British contraceptive collectives to form the National Birth Control Association, later known as the Family Planning Association. This book is an institutional history of the association and its work to scientifically understand, refine and regulate pre-pill contraceptive technology.

The association achieved these goals through a process of standardisation, which ultimately encompassed every aspect of its influence and activity. This ranged from individual contraceptive products, to clinical trials and practices, doctor–patient interactions, record keeping, lay education and therapeutic guidance, and the development and implementation of a medical training syllabus for doctors and nurses. I argue that the association realised that it could exert great socio-cultural, medical and political influence through effective scientific standardisation of its contraceptive policies, products and procedures. Through successive processes of self-regulation, agitation and expansion the NBCA/FPA proved that contraception was ultimately medico-social, as opposed to either medical or social, which the medical profession and governments had always maintained.

As a result of recent and ongoing sexological, medical and scientific advances, the NBCA/FPA and its affiliates, including but not limited to, the Eugenics Society, Birth Control Investigation Committee, and the Society for the Provision of Birth Control Clinics, developed and shared an ideology that supported the scientific legitimisation and standardisation of mechanical and chemical contraceptive technologies. From 1930 the contraceptive lobby united to perfect a replicable scientific methodology that would be acceptable to the medical profession and government regulators. It aimed to transmit the responsibility to provide and guarantee contraceptive safety and efficacy to some more official body or group.

Professionalising Birth Control Through Medico-Scientific Sex Research

As contraceptive historians have routinely demonstrated, the 1920s were an integral decade for the contraceptive movement in Britain and worldwide.[55] Until then, any and all behaviour aiming to curb or increase the chance of potential pregnancy occurred at the discretion of individuals. Birth control, the myriad of socio-sexual practices that individuals, couples and societies applied to limit births, had gained some social acceptance as an ideal, and as a result demand for sex and contraceptive products, methods and knowledge began to intensify and reverberate amongst the middle and lower classes.[56] Contraception was still in many ways sub-cultural, existing outside major spheres of acceptability and candidness, but the principle and techniques became increasingly accessible to couples wanting sexual, contraceptive and/or maternal health guidance. Contraceptive literature and products were disseminated by social hygiene and pseudo-scientific societies, as well as by individuals existing within and around the currently zenith eugenics movement.[57] Prominent advocates on both sides of the Atlantic had independently initiated propaganda efforts to extol and publicise contraceptive knowledge and practice. In the United Kingdom, social reformer Marie Stopes did so through the publication of several prominent, seemingly scientific works on sex and birth control and the foundation of her Society for Constructive Birth Control and Racial Progress clinics, which were quickly complemented by other independent pressure groups that shared her objectives; the Society for the Provision of Birth Control Clinics, Workers Birth Control Group, Birth Control International Information Centre, the Birth Control Investigation Committee and the Eugenics Society.[58] In the United States, nurse and birth control reformer Margaret Sanger founded the Planned Parenthood Federation of America, and worked with the National Committee on Federal Legislation for Birth Control, and the American Birth Control League to take up the task. During the 1920s, the first global formalised and somewhat tolerated attempts to provide contraceptive access to ordinary women via women's welfare/birth control centres were effected; however, ongoing formal, institutional, medical and social opposition to contraceptive practice and principles meant that birth control societies needed to adapt and legitimise the field to expand the population welfare agenda by any practical means.

The predecessor of the British Family Planning Association, the National Birth Control Council (NBCC), which shortly became the National Birth Control Association, was founded in England in 1930. The council incorporated several distinct societies promoting family limitation and child spacing, and by default moral, racial and public health. The council's mandate was to ensure that 'married people may space or limit their families and thus mitigate the evils of ill health and poverty'. A gushing history published in the *Lancet* in 1990 depicts its foundation, early years and agenda:

> In Britain in 1921 when Marie Stopes opened London's first birth control clinic, the average married woman gave birth to eight children... and one woman in a hundred died in child birth. The sad statistics that spurred birth control pioneers were no defence against the scorching verbal attacks of the Church and the medical profession (gynaecologists were among the fiercest opponents) and the more physical displays of disapproval (eggs, bricks and other missiles) from the public. But the courageous campaigners persisted and in 1930 the National Birth Control Council was formed, with 20 clinics. And in 1939 the council changed its name to the Family Planning Association.[59]

This account grossly oversimplifies the progress of societies aiming to provide contraception. Further, the assertion that the 'medical profession', especially gynaecologists en bloc, specifically opposed the provision of contraceptives and family planning education and advice is incorrect, recent nuanced investigations into the topic show a definite gender-based distinction in acceptance and support, at least for specific women who required them on expressly medical grounds.[60] As the century progressed, the profession's opposition and caution were increasingly challenged from within. By 1930 many prominent British contraceptive advocates were prominent physicians and scientists, and often also eugenists.

The association developed its contraceptive agenda in partnership with the local Eugenics Society.[61] Both groups supported birth control and ultimately aimed to discourage excessive births amongst poorer, less healthy citizens; although importantly only the latter group had a pro-eugenic agenda promoting higher birth rates for healthy and wealthy British stock. This relationship encouraged the NBCA to develop a scientific, social fairness and service-based model of contraceptive advice and provision. This position led the association to become the first of two trusted societies publicising and providing birth control in the mid-twentieth century, the other was the International Planned Parenthood Federation.[62]

It was against a backdrop of social and population control philosophies and progress that the local birth control movement emerged and began to garner the publicity and momentum required to make contraception socially and financially known and viable en masse. As this transition occurred, the scientifically aligned eugenic and neo-Malthusian ideologies gradually assumed primacy and through their provable, intervention-based methods moved society, governments and individuals to embrace the social, practical and scientific principles of birth control.

Reaching the goal of mass acceptance and astute and sound employment of contraception required official recognition that 'birth control' was a legitimate practice for couples to employ at their discretion. However, from the 1920s societal mistrust developed towards the term, as it was applied by early proponents such as Marie Stopes and its connections with negative eugenic ideals.[63] This prompted the NBCA's 1939 resolution to abandon the term 'birth control' for 'family planning'; an astute marketing manoeuvre to legitimise and improve official perception of its contraceptive services and methods, and a public relations effort to placate and reassure its ideally poor and grateful clientele that it had their individual interests and family aspirations at heart. In the face of cultural changes that rendered the term 'birth control' outdated and compulsory, this new name reaffirmed the association's long-held pro-fertility attitude and agenda.[64] 'Family planning' would thus make the science of eugenics modern and palatable through inclusive language that championed self-determination via medicine and science. To address concerns around compulsion through prescription, a grey area around informed consent, birth controllers and eugenists engaged interested and sympathetic scientists and physicians to develop and oversee the efficacy, safety and cost-effectiveness of the association's preferred contraceptive techniques and technologies, as well as its marital and sexual hygiene instructions and offerings.[65]

The first quarter of the twentieth century was a time of impressive conceptual and scientific progress, not least in terms of investigating and understanding sex and fertility.[66] Societies espousing 'family planning' perceived the link between their goals and those of scientists investigating erotic practices, proclivities and norms, and sexual, marital and reproductive health, wellbeing and demographics. The time was nigh to accept Iwan Bloch's 1908 challenge to study sexual science in its 'proper subordination as a part of the general "science of mankind"', that being a merger of other major scientific branches: 'of general biology, anthropology and ethnology, philosophy and psychology, the history of literature and the entire history of civilization'.[67] Thus, an interdisciplinary approach to fertility and contraceptive studies and practice became the goal of global contraceptive advocates. It was this new way of thinking about the intersections of sex and medico-science that prompted the introduction of biological and chemical science to contraception in the 1920s.[68]

Simultaneously, two pioneers of contraception, Margaret Sanger in the United States and Marie Stopes in Great Britain, arrived at identical means of legitimising contraception as a health and wellbeing practice.[69] Each observed the 'impact on science of a wave of popular interest in the regulation of human reproduction' and determined to recruit to the cause biologists and chemists who agreed the principles of contraception and science were consonant.[70] This demanded the collection and creation of 'scientific data on which more adequate medical guidance [regarding contraception] could be based'.[71] This would activate a perceived silent but growing minority of medical and scientific professionals who would vocally support the practice when a method was proved to be safe, effective and foolproof.[72] Essentially, to convince the scientific community of the medico-scientific nature of contraception, that truth had to be scientifically proven by birth controllers.

This was achieved by placing contraception within a cannon of acceptable scientific research and knowledge. Specifically, within the extant and increasingly well-regarded domains of sexology and sex research, chemistry and biology.[73]

By the late nineteenth century, science was being established as an exploratory and explanatory pillar of life and in the (incrementally) increasingly secular society, researching the mechanics, proclivities, practices and results of sex was becoming viable. But even the most 'scientific' sex research remains 'constituent of [the] social reality' in which it is generated.[74] So, the research being produced cannot be considered completely and impartially 'scientific', even as a branch of medicine, because that too is laden with popular but not scientifically sound diagnoses and interpretative models. Therefore, the 'life sciences and medicine construct … bodies [of understanding, which would then] give rise to social belief'.[75] Essentially society has and does shape and direct scientific endeavour, which then informs and affects society. Regardless, by the turn of the twentieth century, the medical and scientific professions had become increasingly intertwined with questions of sex, fertility and reproduction—with the exception of contraception and abortion.

So, formalising the former into respectable and reproducible studies became a social and scientific focus for birth controllers and it was here that sympathetic scientists, physicians and philanthropists situated their work. Where sex and contraception lay at the heart of scientific research it was near impossible to create pure, unadulterated knowledge, but this was the task that British (and American) medical contraceptive communities and leaders set for themselves. They sought to situate their work in the cannon of respectable sex research, that included studies of sexual preference and activity, the mechanics of gendered sexual response, the psychology of sexual divergence and 'normal' and 'abnormal' categorisation of sex.[76] This also prompted medical and scientific sex research that fused physiological, psycho-social and lab-based investigations into endocrinology.[77] Sex hormones were both physiologically and demographically important, and chemical and biological sex research become favourably regarded by the mature scientific branches of Physics and Chemistry.[78]

The 1920s was the decade when hormones were appreciated as offering the means to achieve the 'chemical perfectibility of human life'.[79] Knowledge gained during the 'heroic age of reproductive endocrinology', was ultimately adopted by British birth controllers to develop methods of scientific standardisation and to design and perfect new contraceptive technologies.[80]

That decade local British physicians, scientists and social activists with an interest in reproductive, sexual and contraceptive health first joined to investigate and remedy overpopulation, poverty and public health concerns. For individuals and groups with this aim, contraception became a potential panacea to address social, health and moral concerns, but it was acknowledged that very little was known about the functional mechanisms of spermicidal products and insertable technologies, the safety of these practices, and even the actual physiology of sex. The realisation of these limits prompted local physicians and scientists to form the Birth Control Investigation Committee, with the explicit aim of understanding human sex, reproduction and contraception through physiological, medical, chemical, biological and statistical investigations. Within ten years of its formation this group became the research arm of the primary contraceptive collective in the United Kingdom; the NBCA.

Developing Sex and Contraceptive Research in Britain

The slow but definite successes of laboratory-based sexual science led contraceptive lobbyists hoping to legitimise the cause and methodology to petition for scientific and medical support to investigate contraception as a proposed solution to address a slew of local and global social, economic, legal, political, health and interpersonal problems.

In 1927 Margaret Sanger convened the World Population Conference in Geneva. The meeting had two prime goals: to discover if science had the potential to arrest disturbing recent population trends (overpopulation in certain regions, and the risk of zero population growth elsewhere), and to encourage the international scientific community to advance 'intelligent solution[s]' to these problems.[81] 'Science' was then emerging from a critical debate regarding its primary function: was it the establishment of knowledge and truth, or the refinement of scientific methods and technologies for public application?[82] The latter, 'applied' science, was usurped as the primary focus of scientific labour and became subordinate to 'pure' scientific enquiry.[83] This conference, with an overtly applied scientific agenda, facilitated the formation of two transnational organisations dedicated to the medico-scientific investigation and promotion of birth control technologies and techniques. The International Medical Group for the Investigation of Birth Control and the International Union for the Scientific Investigation of Population Problems shared goals to 'study the various problems of population' and provide authoritative 'instruction in the most satisfactory method of contraception'.[84] Each also sought to commandeer scientific methodology to legitimise the practice and methods of birth control. But this caused a serious dilemma: how to develop a field of scientific enquiry that would be tolerable to the mature schools of science dedicated to 'pure' knowledge, that could also be 'applied' and tested within contraceptive clinics for public and individual benefit?

During the early twentieth century neither science nor medicine co-existed harmoniously with contraception.[85] Most physicians, scientists and politicians were wary of the social and professional risk of allying themselves with 'the controversial question of birth control'. Physician and eugenist Carlos Patton Blacker explained in 1928 that even professionals that did not oppose the morality of contraception, viewed it 'with apprehension as being racially harmful'.[86] Some scientists argued the tactical merger between the eugenics and birth control movements was unhelpful to the latter cause; diluting its altruistic potential through affiliation with loaded medico-morality that proposed to remedy social problems by focusing fertility limitation on the poorest and most vulnerable sectors of the public. There was a risk that the contraceptive movement might be burdened by its connection with the socio-utilitarian eugenics movement, and this threatened to derail its proposed alliance with scientific and medical communities. So, the nature of the relationship was significantly checked, although the philanthropic ideological fundamentals of the partnership remained. Sanger's planned 'solely scientific' conference on population control, downplayed the fact that the birth control movement was more a social movement than a scientific one by stressing that it was 'thoroughly rational from a scientific point of view'.[87]

The British eugenic community were frank about their socio-political agenda which aimed to 'spread the practice of birth control'; but acknowledged there was 'general mistrust of the efficacy of [extant] contraceptive' methods and technologies and promised that it would work with local birth control advocates to scientifically develop and standardise an effective and trustworthy method to achieve its political and demographic goals.[88]

The conveners of the World Population Conference, Sanger, Professors Julian Huxley, A. M. Carr-Saunders, and E. M. East, Doctors F. A. E. Crew and C. C. Little, and financier Clinton Chance, were adamant the summit was integral to contraception being recognised and the foundation of contraceptive science as a legitimate field of study. They were keen to use the conference to 'piece together' the current state of knowledge and work being undertaken, and 'quietly and efficiently make [sympathetic scientific] contacts'.[89] Thus, attendance was 'strictly limited to persons of established scientific standing', with most attendees hailing from birth control clinics and collectives in the United States, Great Britain and the countries of Western Europe, with a few outliers from East Asia, South America and Australia.[90] Despite organisers' insistence that 'applied questions' of 'birth control or Neo-Malthusianism shall not appear as being a dominant element' of the event, contraception was tackled as a budding branch of both pure and applied scientific and medical enquiry fusing physiology, biology, chemistry, medicine and statistics.[91] The organisers used the conference to manoeuvre the formation of a branch of pure contraceptive science under a less politically and socially perilous title: 'population science'.[92] The summit was a spectacular success in uniting scientists and physicians interested in contraception 'in a true scientific spirit [to] discuss these great controversial questions' regarding the form, method and object of contraceptive technologies, and how to answer them through scientific investigation.[93]

In early 1928, at 'the initiative of lay members of the North Kensington and Cambridge Birth Control Clinics', Margery Spring Rice and Lella Florence, wealthy patrons who had taken the provision of scientific birth control as their political and philanthropic focus,[94] several conference attendees convened a new committee of experts to investigate scientific and medical aspects of contraception with 'neutrality and impartiality'.[95] The Birth Control Investigation Committee was led by Professor Sir Humphrey Rollinson, former President of the London Medical Society, Royal Society of Medicine and Royal College of Physicians and current Physician-in-ordinary to the King. He was joined by Blacker, Carr-Saunders, Crew, surgeon Mr. C. J. Bond, prominent endocrinologist Dr. Francis H. A. Marshall, electrophysiologist and 1932 joint Nobel Prize for Physiology recipient Professor Edgar Adrian, evolutionary biologist and eugenist Professor Julian Huxley, physician and chemist Professor Arthur Ellis, practising and academic physiologist Professor Winifred Cullis and medical officer at the Walworth Women's Welfare Clinic Dr. Gladys Cox. Later they were joined by physician, fertility researcher and birth controller Margaret Jackson and gynaecologist and educator Helena Wright. This prestigious medical and scientific collective aimed to address the fact that 'contraception is widely used... [but] there is very little medical and scientific knowledge about it'.[96] At its inception the committee outlined research programs integral to understanding attitudes towards contraception, its technologies and methods, and ensuring quality and safety.

Within months, the committee established two scientific sub-committees: Statistics and Research. The statistics sub-committee would collect, analyse and distribute scientific data collected from local and national contraceptive clinics, and the research sub-committee was charged with 'defin[ing] problems on which investigation is necessary,... order[ing] their urgency', and instigating laboratory and clinical investigations in order to uncover solutions to contraceptive problems.[97]

Through this inoffensive structure the group defined two distinct approaches to the 'dispassionate investigation of the subject of birth control'. A 'sociological or applied' approach and a method that was 'purely scientific'.[98]

This agenda was effectively devised to straddle both sides of the ongoing conflict about the value and use of scientific enquiry. It would generate pure scientific contraceptive knowledge in addition to undertaking 'applied' research for public benefit. Contraception was a unique product in the context of this debate as its potential utility and risks had the capacity to fundamentally change social, religious, sexual, political and cultural structures and norms, hence it had to be navigated with extreme caution.[99] For the BCIC, this meant that delineating the margin separating its 'applied' and 'pure' activities was, at least initially, extremely important and could not be infringed. The group maintained that only the work undertaken by its statistics sub-committee would address applied scientific contraception.[100] This work was definitively sociological, requiring mathematical assessment and analysis of contraceptive clinic patient 'case cards' and responses to strategically composed and disseminated questionnaires to reach conclusions regarding the

> relative merits, in actual use, of the various birth control methods now employed ...whether one method ... in the hands of different clinics or different doctors, can vary its efficiency and success ... [and whether contraceptive] methods are equally valuable for members of different social and economic classes.

These findings had the potential for immediate application for the benefit of patients and community education. Further statistics sub-committee research proposed to use these methods to ascertain if contraception had any psychological or physiological impact or compromised future fertility.[101] This work was carried into the daily operations and political and medical agendas at centres run by the Society for the Provision of Birth Control Clinics and Workers Birth Control Group.

Only laboratory-based work was deemed 'pure' and the BCIC made sure to place this research agenda definitively within the spectrum of the 'laborious nature of [historical] scientific research'. Thus, the committee's initial primary research focus, spermicides and spermatoxins, were framed as the result of the 'long pioneer researches of Pasteur, Erlich, Metchnikoff and their thousands of followers'. This strategy to legitimise the foundation of this emerging scientific field as the inevitable result of some of the greatest achievements in science and medicine, validated the BCIC's efforts to uncover the 'purely scientific basis of birth control'.[102] It was of course understood to be entirely inevitable that practical utility for this research would be discovered, but this was initially downplayed to stress the absolute necessity of this base knowledge.

In 1930 a statement detailing BCIC-directed 'pure' research was devised. This included research into the 'physiology of coition', as the scientific and medical profession realised it was 'surprisingly ignorant of the detailed physiology of the sexual act'.[103] This research aimed to discover if 'during coitus, the uterus behaves in such a way as actively to suck up the contents of the vaginal vault'.[104] Discovering this was imperative, as it would impact the dependability of contraceptives, and necessitate extensive safety testing if chemical contraceptives were 'aspirated' into the uterus upon orgasm. Animal trials injecting chemical contraceptives into rabbit and dog uteri to assess risk commenced[105]; later human radiological trials were undertaken to definitively determine suction.[106] Further research investigated the 'chemistry of the female genital tract', specifically fluctuations of vaginal acidity, PH levels, cytology and flora during the menstrual cycle.[107] A further initial line of enquiry was hormonal. The BCIC realised it was 'quite likely … further knowledge [would] make it possible to achieve temporary sterility by injecting certain hormones, or even administering them by mouth'. However, despite the committee funding a series of experiments to understand reproductive hormonal systems (a novel and entirely necessary interwar pursuit), the group was never in a position to undertake 'pure' or 'applied' research in this area. Contraception via high temperatures, x-ray and 'mechanical methods' were also considered,[108] but only the latter was formally explored in relation to intra-uterine devices.[109] This work was undertaken by recent recruit, Wright, who led the association's medical, scientific and educational pursuits for three decades.[110]

Emphasising laboratory research constitutes the first stage in the 'domestication process' of a scientific field.[111] The BCIC certainly attempted to tap into this site of 'pure science' to carve itself a laboratory-based niche wherein it carried out and broadcast its work investigating sex and contraception. Meanwhile its 'sociological or applied' research was firmly confined within affiliated clinics and was, at least publicly, clearly separated from the laboratory, and hence the committee's pure scientific endeavours.[112] Caroline Rusterholz has shrewdly identified the gendered distinction that ran through the BCIC's delineation of its research activities and agenda, and their concomitant importance and legitimacy within the broader realms of science and medicine; men, specifically three University-based researchers Harry Carleton, John Baker and Cecil Voge, worked in laboratories tackling only 'pure' scientific and medical questions, and women doctors and lay staff in the already questionable contraceptive clinics were charged with undertaking the applied scientific pursuits.[113] This gendered lens can certainly help to understand how and why the scientific legitimacy of contraception took so long to establish, even in the wake of these early aspirations and wide-ranging activities; the application of contraception was ultimately perceived, however inaccurately, to be gendered female and so, there was little urgency for wider male professional action in this budding field.

The separation of the BCIC's 'pure' and 'applied' research was short lived. Tensions centring on the 'unsatisfactoriness of present methods' arose quickly and members began to worry that the committee's pure investigations 'might prove harmful to the birth control movement' and that its scientific findings ought to benefit the nation and its people.[114]

This revelation coincided with the inception of the National Birth Control Council in July 1930 when Spring Rice the founder of the SPBCC affiliated North Kensington Women's Welfare Clinic implored the society to 'consider the necessity of forming a central organisation' for administering and overseeing birth control technologies.[115] This coalition merged contemporary British contraceptive pressure groups, the SPBCC, Society for Constructive Birth Control and Racial Progress and Birth Control International Information Centre, into a bigger and more powerful organisation with further reaching services and better lobbying potential.[116]

The BCIC joined only after 1930 when it accepted it had failed to achieve its goal of creating a school of pure contraceptive inquiry distinct from its application and amended its original constitution to include both pure and applied scientific goals. It would now

(a) Organise research with a view to discovering better [contraceptive] methods.
(b) Investigate existing methods in order to ascertain to what extent and with what results they are practised.[117]

In June 1931 the BCIC agreed to 'association and co-operation with the NBCC'. The following month, the collective officially became the National Birth Control Association, 'a central organisation' for administering and overseeing contraception in Britain.[118] The BCIC became a 'Special Committee' that boasted the autonomy to self-direct scientific contraceptive research and development. Thereafter an applied science focus of enquiry characterised BCIC and NBCA work to 'advocate and promote the provision of facilities for scientific contraception' to married Britons.[119]

Navigating Eugenic and Contraceptive Science

It would be remiss of me to construct a history of the NBCA/FPA as a medico-scientific organisation without attending its somewhat symbiotic relationships with the local Eugenics Society and its ideologies and methodologies. But it would also be wrong to overplay the connection as the groups were never really ideologically aligned, even if they boasted some shared interests, aspirations and members. This section briefly describes the nature of the relationship between the pair and their fluctuating flirtations with science, legitimacy, medicine, sex and contraception, education and potential collaboration before the 1940s.

Early in its life, the NBCA's research and didactic focus centred exclusively on clinic-based instruction on contraceptive methods that had been scientifically proven effective to justify the practice and persuade opponents that modern contraceptive technologies were safe and efficacious. Eugenic work never officially fell within the immediate purview of NBCA/FPA medical or scientific staff, nor was it formally located in its clinics, literature or methodology.[120] The NBCA and society worked closely to ensure public access to practical and accessible products and information on fertility and contraception; this was achieved despite the groups fundamentally disagreeing on the desired social applications of developing contraceptive science and technology. Each group was committed to discovering and applying scientific truths about contraception to produce stronger children in stable and prepared families. But they disagreed on the role of eugenic science within the association's national contraceptive agenda. From 1930 to the mid-1940s the groups collaborated closely to legitimise pure and applied sex and contraceptive science for medical, scientific, educational and governmental oversight. This was effected through the cultivation of support for scientific marital, sex, contraceptive (and to some degree, eugenic) knowledge.

In 1930, Eugenics, the broad 'study of the laws of human life', 'human heredity and the conservation, evolution and progress of the human race', was ascendant and well regarded.[121] The local society was peopled by many prolific scientists and physicians, including Blacker and Rollinson who straddled each of the organisations under discussion. In 1930 it boasted over 500 members and subscribers and used its significant funds to support its scientific, educational and lobbying goals and those of its friends.

At its core, the Eugenics Society was propagandist, policy-oriented and mercenary when it came to contraception and its applications. Its scientific endeavours in the field were fundamentally focused upon application and outcome. Its grants of funds for scientific research into contraception had to have the potential for application, specifically by 'low grade persons', and prevent overreliance on methods that 'demanded a certain level of intelligence and dexterity', as these were the most likely to fail in use. To these ends it funded pure research and development with the BCIC: specifically, spermicidal analyses and the design of a foolproof chemical contraceptive.[122]

The society was dedicated to technological intervention that would ensure the infertility of those suffering from 'mental or biological weakness'. Eugenically driven contraceptive intervention was envisioned and promoted, without apparent support from the NBCA.[123] Eugenists long believed that contraception, though generally good, had two limitations: it was too frequently employed by those likely to raise 'promising families', and too little by those perpetuating 'problem families'.[124]

The Eugenics Society worked to educate the public, medical profession and political classes of the benefits of sterilisation to elicit support and encourage favourable legislation.[125] But, there was little support for eugenic technologies of sterilisation in interwar Britain.[126] The NBCA also decided voluntary sterilisation was not its objective. But there was likely some in-principle agreement on the need for eugenic contraception and that different groups required radically different contraceptive interventions. Addressing the needs of those with the least capacity and dexterity, led the association to hone its scientific and practical focus on discovering and promoting the most simple, safe and effective family planning methods.[127]

The NBCA and Eugenics Society maintained a complex ideological, fiscal and political relationship throughout the 1930s. Their interwar relationship was intimate and symbiotic as each bolstered the other's socio-political agenda through research and promotional work, and they collaborated on scientific, medical and educational projects spearheaded by the BCIC.

In 1936 the groups perceived a merger might be mutually advantageous.[128] It was proposed that the society would become the parent and the NBCA would be reconstituted as the Birth Control Committee of the Council of the Eugenics Society. In return the society would fully fund all NBCA activities and define and provide a standardised education program (from a eugenic and marriage guidance perspective) for application in all association clinics. This merger would occur incrementally as opposition was anticipated owing to some 'misapprehension of the aims and political orientation of the Eugenics Society'.[129]

One of the main issues of contention during discussions regarded the application of contraception. The NBCA wanted contraceptive education and methods to be freely available to all[130]; whilst the Eugenics Society maintained that contraceptive application should be limited as it had the capacity to diminish the 'differential birth rate'. There was also apprehension about the concern that may be caused in the public mind by linking 'the many controversial matters' in the eugenic doctrine with clinic-based contraception. Before the association would agree to merge, it requested the society change its name and 'drastically' update its 'Aims and Objects'.[131] Outside of these areas of contention, the groups agreed.

Unfortunately, these issues proved unresolvable and discussions concluded without agreement, but the groups remained closely allied. In 1937 the central NBCA office was relocated to the Eugenics Society premises (at a very favourable rate),[132] to expedite their shared social, political and scientific contraceptive agendas.[133] The groups jointly argued that only comprehensive, accessible and standardised contraceptive methods and education could prevent future generations from unnecessary suffering.

In early 1937 the groups amassed a delegation representing eleven societies concerned with contraception and maternal and child health that persuaded the Ministry of Health to issue *Circular 1622*[134]; it stressed the necessity for post- as well as antenatal services in local health authority clinics and specifically supported the provision of 'contraceptive advice to women attending the clinic' for whom further pregnancy could be detrimental.[135] A shared victory and proof of their ongoing ideological and pragmatic goals.

****

This work is primarily based upon archival material held in the Wellcome Library in London, United Kingdom. The archives that have been used to build this argument are almost exclusively those within the Family Planning Association archive, which is a collection of materials relating to the FPA and its predecessor bodies between 1921 and 1976. These include papers relating to FPA administration, general council and subcommittee meeting records, conferences, contraceptive testing, training and education activities, research and surveys, branches, advertising and publicity, and important figures, such as Margery Spring Rice and Caspar Brooks. Further, I heavily utilise the papers of the Eugenics Society and its prime activist and educator Carlos Patton Blacker. My research exploits an underutilised selection of materials in these archives which record the day-to-day management of the association and its clinics and focuses on setting standards for practice and products. This aim is often not explicit in the archival material, which may go some way to explaining its lack of previous attention but is an ever-present focus throughout the FPA early records and those of its collaborators.

Other archival material that has been utilised is held in the University of Melbourne Archives. The Victor Hugo Wallace Papers are a collection of personal and professional papers from Melbourne's prime birth control activist in the early to mid-twentieth century. These papers relate to Wallace's private practice, research, membership in the Eugenics Society of Victoria, and educational activities relating to contraception and marital and sexual relations. I have also accessed online archives from the Countway Library of Medicine in Boston, Massachusetts regarding the scientific investigations of Clarence Gamble relating to spermicidal effectiveness, and the National Committee on Maternal Health and its activities promoting medicalised contraception. Finally, the online collections of the Bureau of Social Hygiene at the Rockefeller Archive Centre in New York were accessed to research the sexological testing activities, which were funded by the Office of Messers. Rockefeller.

****

Chapter 2 takes the North Kensington Women' Welfare Clinic as its focus to consider the development of contraceptive clinic form, function and practice standardisation under FPA management and direction. It challenges the common perception that clinic operation functions as a historical known-unknown by detailing the various medical and management pursuits that were pioneered and exported from the site in order to link the clinic with medicine and science in public perception. Through a case study of the site's medico-scientific standards-focused management and its therapeutic and educational work, I illuminate how contraceptive and administrative technologies were understood, prescribed and used therein to encourage a public perception of the alignment between birth control methods and practices and necessary national preventive and maternal health goals.

The two major standardisation foci that the association identified from its trial clinic in North Kensington are the focus of the subsequent two chapters. Chapter 3 details the association's efforts (as an organisation and the work of some of its individual members) to regulate and disseminate scientifically true, standardised contraceptive and sex education with the assistance and cooperation of family-focused societies, namely the Eugenics Society, National Marriage Guidance Council, the Family Relations Group and the Central Council for Health Education. Through the concept of the development of knowledge networks it charts the shift away from female-centric, personalised, in-clinic education and advice, towards the NBCA/FPA's stated goal of developing a state- and medical profession-implemented and regulated contraceptive syllabus, geared towards providing adequate scientific, medical, and moral advice—for both medical professionals and lay people.

Chapter 4 details the efforts of the association and its US counterparts, to commission British scientists to develop a universally applicable qualitative laboratory-based technique to assess the efficacy, quality and safety of extant chemical technologies, and also their parallel efforts to discover or develop a scientifically verifiable perfect single-use contraceptive. Through these tests, the NBCA/FPA developed and published an annual approved list of contraceptives, touted as a definitive register of effective contraceptive devices and compounds sold in the United Kingdom. It also charts the association's work to collect and exploit clinic data to understand sociological aspects of contraceptive use including convenience, ease of application, acceptability to both partners and to discover the causes of failures of contraceptive technologies and of patients to return to NBCA/FPA clinics. All these medico-scientific research activities were applied to convince those British contraceptive manufacturers that desired to have their products prescribed and sold in clinics, that gaining NBCA/FPA approval, by refining products to meet its testing and acceptability standards, would be fiscally practical, as the association increasingly dominated the market through its contraceptive clinics, and worked to become the arbiter of British contraceptive products and advice.

The concluding discussion in Chapter 5 examines the expansion of the association's efficacy, safety and quality standardisation in the era of the pill. It describes and charts the NBCA/FPA's British research and regulatory networks that merged to bring the association's scientifically verified product standards, teachings and clinic practices under the legislative and professional remit of medical, educational and government-allied bodies. By detailing the FPA's efforts to attract official regulators like the British Standards Institution and British Pharmacopeia to overtake its regulatory work, I argue that the association was finally ready to hand over its standards work to groups that now agreed that science was the correct arbiter of contraceptive methods and products. This was also the case for the medical profession, but in order for it to finally accept a duty to patients regarding family planning and limitation, two major technological and bureaucratic changes were needed: the emergence of a definitely medical birth control technology, the hormonal oral contraceptive, and the foundation of a specialised research body to independently and impartially assess and test contraceptive methods and products, the Council for the Investigation of Fertility Control. It charts the FPA's contributions to the development and implementation of small- and large-scale human clinical and acceptability trials for the pill, and its efforts to exploit its hard-earned scientific and medical authority to advise and direct its dedicated contraceptive consumer base, to stagger the release of oral contraceptives, until it, in direct collaboration with the council, had developed and confirmed independent scientific means to verify and regulate the product, in accordance with the standards it applied to all other contraceptives issued at its clinics.

This book explores how the NBCA/FPA and its affiliates utilised and manipulated scientific, medical, technological, educational, social and legislative avenues to concurrently legitimise and standardise the clinical practice and technologies of contraception available in Britain in the mid-twentieth century.

Notes

1. C. F. Chance, Birth Control Investigation Committee Draft Goals, 2 March 1928. Wellcome Library, Archives of the Eugenics Society (WL/SA/EUG), WI/SA/EUG/D/12.
2. Merriley Borell, 'Biologists and the Promotion of Birth Control Research, 1918–1938', *Journal of the History of Biology* 20, 1 (1987): 51–87; Ilana Löwy, '"Sexual Chemistry" Before the Pill: Science, Industry and Chemical Contraceptives, 1920–1960', *British Journal for the History of Science* 44, 2 (2011): 245–74; Richard Soloway, 'The "Perfect Contraceptive": Eugenics and Birth Control Research in Britain and America in the Interwar Years', *Journal of Contemporary History* 30, 4 (1995): 637–64; Richard Soloway, *Demography and Degeneration: Eugenics and the Declining Birthrate in Twentieth Century Britain* (Chapel Hill: University of North Carolina Press, 1995); Adele E. Clarke, *Disciplining Reproduction: Modernity, American Life Sciences, and the 'Problems of Sex'* (Los Angeles: University of California Press, 1998); Caroline Rusterholz, 'Testing the Gräfenberg Ring in Interwar Britain: Norman Haire, Helena Wright, and the Debate Over Statistical Evidence, Side Effects, and Intra-uterine Contraception', *Journal of the History of Medicine and Allied Sciences* 72, 4 (2017): 448–67.
3. P. Conrad, 'Medicalisation and Social Control', *Annual Review of Sociology* 18, 1 (1992): 211.
4. H. Marland, 'Women, Health and Medicine', in *The Oxford Handbook on the History of Medicine*, ed. M. Jackson (Oxford: Oxford University Press, 2011), 484–502.
5. Michel Foucault, *The History of Sexuality, Vol. 1: An Introduction* (New York: Pantheon Books, 1978).
6. Lara Marks, *Sexual Chemistry: A History of the Contraceptive Pill* (London: Yale University Press, 2001), 116–8.
7. Borell, 'Birth Control Research', 51–87; Löwy, 'Sexual Chemistry', 245–74; Soloway, 'The "Perfect Contraceptive"', 637–64; Clarke, *Disciplining Reproduction*; Lara Marks, *Sexual Chemistry*; Bernard Asbell,

The Pill: A Biography of the Drug that Changed the World (New York: Random House, 1995).

8. Caroline Rusterholz, *Women's Medicine: Sex, Family Planning and British Female Doctors in Transnational Perspective, 1920–70* (Manchester: Manchester University Press, 2020).
9. Simon Szreter and Kate Fisher, *Sex Before the Sexual Revolution: Intimate Life in England 1918–1963* (Cambridge: Cambridge University Press, 2010); Rose Holz, *The Birth Control Clinic in a Marketplace World* (Woodbridge: Boydell & Brewer, University of Rochester Press, 2012); Cathy Moran Hajo, *Birth Control on Main Street: Organizing Clinics in the United States, 1916–1939* (Urbana: University of Illinois Press, 2010); Lara Marks, *Metropolitan Maternity: Maternal and Infant Welfare Services in Twentieth Century London* (The Netherlands: Rodophi B. V. Amsterdam, 1996); Marks, *Sexual Chemistry*; Audrey Leathard, *Fight for Family Planning: The Development of Family Planning Services in Britain, 1921–74* (London: The Macmillan Press, 1980); Hera Cook, *The Long Sexual Revolution: English Women, Sex and Contraception, 1800–1975* (New York: Oxford University Press, 2004); Beryl Suitters, *Be Brave and Angry: Chronicles of the International Planned Parenthood Federation* (London: International Planned Parenthood Federation, 1973); Clare Debenham, 'Grassroots Feminism: A Study of the Campaign of the Society for the Provision of Birth Control Clinics, 1924–1938' (PhD, University of Manchester, 2010); Caroline Walker, 'Making Birth Control Respectable: The Society for Constructive Birth Control and Racial Progress, and the American Birth Control League, in Comparative Perspective, 1921–1938' (PhD, University of Bristol, 2007); Soloway, *Demography and Degeneration*.
10. Deborah A. Cohen 'Private Lives in Public Spaces: Marie Stopes, the Mothers' Clinics and the Practice of Contraception', *History Workshop* 35 (Spring 1993): 399–411; Kate Fisher, 'Contrasting Cultures of Contraception: Birth Control Clinics and the Working-Classes in Britain between the Wars', in *Biographies of Remedies: Drugs, Medicines and Contraceptives in Dutch and Anglo-American Healing Cultures*, eds. Marijke Gijswijt-Hofstra, G. M. van Heteren and Tilli Tansey (Amsterdam: Rodopi, 2002), 141–57.
11. Ann Collier, *The Humble Little Condom: A History* (New York: Prometheus Books, 2007); Chikako Takeshita, *The Global Biopolitics of the IUD: How Science Constructs Contraceptive Users and Women's Bodies* (Cambridge, MA: MIT Press, 2012), 42.
12. Jesse Olszynko-Gryn, 'The Demand for Pregnancy Testing: The Aschheim-Zondek Reaction, Diagnostic Versatility, and Laboratory Services in 1930s Britain', *Studies in History and Philosophy of Science*

Part C: Studies in History and Philosophy of Biological and Biomedical Sciences 47, B (2014): 233–47; Jesse Olszynko-Gryn and Caroline Rusterholz, 'Reproductive Politics in Twentieth-century France and Britain', *Medical History* 63, 2 (2019): 117–33; Jesse Olszynko-Gryn, 'Laparoscopy as a Technology of Population Control: A Use-Centered History of Surgical Sterilization,' in *A World of Populations: Transnational Perspectives on Demography in the Twentieth Century*, eds. Heinrich Hartmann and Corinna R. Unger (New York: Berghahn Books, 2014), 147–77; Jesse Olszynko-Gryn, 'Contraceptive Technologies', in *Twentieth Century Population Thinking: A Critical Reader in Primary Sources, The Population Knowledge Network* (London: 2015), 172–208; Jesse Olszynko-Gryn, 'Technologies of Contraception and Abortion', in *Reproduction: Antiquity to the Present Day*, eds. Nick Hopwood, Rebecca Flemming and Laura Kassell (Cambridge: Cambridge University Press, 2018), 535–51; Claire Jones, '"Under the Covers?" Commerce, Contraceptives and Consumers in England and Wales, 1880–1960', *Social History of Medicine* 29, 4 (2016): 734–56; John Peel, 'The Manufacture and Retailing of Contraceptives in England', *Population Studies* 17, 2 (1963): 113–25; John Peel, 'Contraception and the Medical Profession', *Population Studies* 18, 2 (1964): 133–45; Ben Mechen, '"Closer Together": Durex Condoms and Contraceptive Consumerism in 1970s Britain', in *Perceptions of Pregnancy from the Seventeenth to the Twentieth Century,* eds. Jennifer Evans and Ciara Meehan (Basingstoke: Palgrave, 2016), 213–60; Jessica Borge, 'Propagating Progress and Circumventing Harm: Reconciling References to Contraceptives in British Television and Cinema of the 1960s', in *Reproductive Rights Issues in Popular Media: International Perspectives*, eds. Waltraud Maierhofer and Beth Widmaier Capo (North Carolina: McFarland and Company, Inc., Publishers, 2017), 11–28; Jessica Borge, *Protective Practices: The London Rubber Company and the Condom Business* (Montreal & Kingston: McGill-Queen's University Press, 2020); Claire L. Jones, *The Business of Birth Control* (Manchester: Manchester University Press, 2020).

13. Asbell, *The Pill*; Elaine Tyler May, *America and the Pill: A History of Promise, Peril and Liberation* (New York: Basic Books, 2010); Marks, *Sexual Chemistry*; Margaret Sanger, *Margaret Sanger: An Autobiography* (New York: Maxwell Reprint Co, 1970); Jean H. Baker, *Margaret Sanger: A Life of Passion* (New York: Hill and Wang, 2011); Vicki Cox, *Margaret Sanger: Rebel for Women's Rights* (United States of America: Chelsea House Publishers, 2005); Heather Munroe Prescott, *The Morning After: A History of Emergency Contraception in the United States* (United States of America: Rutgers, 2011); Armond Fields, *Katharine Dexter McCormick: Pioneer for Women's Rights* (United States

of America: Praeger, 2003); Carl Djerassi, *This Man's Pill: Reflections on the 50th Birthday of the Pill* (Oxford: Oxford University Press, 2001); Shirley Green, *The Curious History of Contraception* (London: Ebury Press, 1971); Margaret Marsh and Wanda Ronner, *The Fertility Doctor: John Rock and the Reproductive Revolution* (Baltimore: Johns Hopkins University Press, 2008); Leon Speroff, *A Good Man: Gregory Goodwin Pincus: the Man, his Story, the Birth Control Pill* (Portland: Arnica Publishing, 2009).

14. Suzanne White Junod and Lara Marks, 'Women's Trials: The Approval of the First Oral Contraceptive Pill in the United States and Great Britain', *Journal of the History of Medicine* 57 (2002): 117–60.
15. Asbell, *The Pill*, 5.
16. Elizabeth Siegel Watkins, *On the Pill: A Social History of Oral Contraceptives, 1950–1970* (Baltimore: Johns Hopkins University Press, 1998), 1.
17. May, *America and the Pill*, 6.
18. 'The Age of the Thing', *Economist*, 25 December 1993: 48.
19. Beryl Suitters, *The History of Contraceptives* (London: The Fanfare Press Ltd., 1967), 20–2; Suitters, *Be Brave and Angry.*
20. Soloway, 'The "Perfect Contraceptive"'; Soloway, *Demography and Degeneration.*
21. Löwy, "Sexual Chemistry", 255–6.
22. Borell, 'Birth Control Research', 81–2.
23. Clarke, *Disciplining Reproduction.*
24. Clarke, *Disciplining Reproduction.*
25. Borell, 'Birth Control Research', 85.
26. Borge, *Protective Practices*; Jones, *The Business of Birth Control*; Jones, '"Under the Covers?"'.
27. Rusterholz, *Women's Medicine.*
28. Clare Debenham, *Birth Control and the Rights of Women: Post Suffrage Feminism in the Twentieth Century* (London: I.B. Taurus, 2013); Debenham, 'Grassroots Feminism', 151.
29. Caroline Rusterholz, '"You Can't Dismiss that as Being Less Happy, You See it is Different". Sexual Counselling in 1950s England', *Twentieth Century British History* 30, 3 (2019): 375–98; Jessica Borge, *The Psychosexual Counselling Tapes of Joan Malleson: New Theories* (MA, University of London, 2012).
30. Szreter and Fisher, *Sex Before the Sexual Revolution.*
31. Thomas Malthus, *An Essay on the Principle of Population, Volume 1* (New York: Cosmo Classics, 2007), 13–4.
32. Brian Dolan, 'Introduction: Malthusian Selections', in *Malthus, Medicine and Morality: 'Malthusianism' After 1798*, ed. Brian Dolan (Amsterdam: Editions Rodopi B. V., 2000), 3.

33. F. D'arcy, 'The Malthusian League and the Resistance to Birth Control Propaganda in Late Victorian Britain', *Population Studies* 31, 3 (1977): 438–9; Eric Neumayer, 'An Empirical Test of a Neo-Malthusian Theory of Fertility Change', *Population and Environment* 27, 4 (2006): 327–9.
34. Charles Darwin, *The Origin of Species* (New York: P. F. Collier and Son Company, 1909), 297.
35. Joël Roucloux, 'Can Democracy Survive the Disgust of Man for Man? From Social Darwinism to Eugenics', *Diogenes* 49, 195 (2002): 48.
36. Marque-Luisa Miringoff, 'The Impact of Population Policy Upon Social Welfare', *Social Service Review* 54, 3 (1980): 302.
37. Miringoff, 'The Impact of Population Policy Upon Social Welfare', 308–10.
38. Roy Porter, ed., *The Cambridge Illustrated History of Medicine* (Cambridge: Cambridge University Press, 1996), 326.
39. Francis Galton, 'Hereditary Improvement', *Fraser's Magazine*, February 1873: 116; Richard Soloway, *Birth Control and the Population Question in England, 1877–1930* (Chapel Hill: The University of North Carolina Press, 1982), 35; Francis Galton, *Inquiries into Human Faculty and Its Development* (London: Macmillan, 1883), 24.
40. Soloway, *Demography and Degeneration*, 362.
41. Soloway, *Demography and Degeneration*, 87.
42. Szreter and Fisher, *Sex Before the Sexual Revolution*, 229–67.
43. Charles Knowlton, *Fruits of Philosophy, Second Edition, with an Introduction by Charles Bradlaugh and Annie Besant* (London: Freethought Publishing Company, 1877), 43.
44. J. A. Banks and Olive Banks, 'The Bradlaugh-Besant Trial and the English Newspapers', *Population Studies* 8, 1 (1954): 28–9.
45. Rosanna Ledbetter, *A History of the Malthusian League: 1877–1927* (Columbus: Ohio State University Press, 1976), 69; Mohan Rao, *From Population Control to Reproductive Health: Malthusian Arithmetic* (India: Sage, 2004), 101.
46. George Drysdale, *The Elements of Social Science: Or Physical, Sexual and Natural Religion, Fourth Edition* (London: E. Truelove, 1861), 349; Helenn Blackman, 'Reproduction Since 1750', in *The Routledge History of Sex and the Body, 1500 to the Present*, eds. Sarah Toulalan and Kate Fisher (United Kingdom: Routledge, 2013), 379.
47. Staffan Müller-Wille and Hans-Jörg Rheinberger, *A Cultural History of Heredity* (Chicago: Chicago University Press, 2012), 98.
48. Francis Galton, *Essays in Eugenics* (London: The Eugenics Education Society, 1909), 65–6; Angus McLaren, *Birth Control in Nineteenth Century England* (Great Britain: Redwood Burn Ltd, 1978), 145.
49. Clyde Chitty, *Eugenics, Race and Intelligence in Education* (London, Continuum International Publishing, 2009), 46.

50. G. R. Searle, ed., *Eugenics and Politics in Britain, 1900–1914* (Leyden: Noordhoff International Politics, 1976), 100.
51. *Pamphlet 'An Outline of a Practical Eugenic Policy'*, [c.1925]. WL/SA/EUG/J/17.
52. John Loadman, *Tears of the Tree: The Story of Rubber—A Modern Marvel* (Oxford: Oxford University Press, 2005), 293–311.
53. Borell, 'Birth Control Research'; Löwy, "Sexual Chemistry"; Soloway, 'The "Perfect Contraceptive"'; Natasha Szuhan, 'Sex in the Laboratory: The Family Planning Association and Contraceptive Science in Britain, 1929–1959', *British Journal for the History of Science* 51, 3 (2018): 487–510.
54. Leathard, *Fight for Family Planning*, 11, 17.
55. Leathard, *Fight for Family Planning*; Suitters, *The History of Contraceptives*; Suitters, *Be Brave and Angry*; Marks, *Sexual Chemistry*; Debenham, *Birth Control and the Rights of Women*; Debenham, 'Grassroots Feminism'; Walker, 'Making Birth Control Respectable'; Borell, 'Birth Control Research'; Löwy, 'Sexual Chemistry'; Soloway, 'The "Perfect Contraceptive"'; Soloway, *Demography and Degeneration*; Clarke, *Disciplining Reproduction*; Ann Farmer, *By Their Fruits: Eugenics, Population Control, and the Abortion Campaign* (Washington: Catholic University of America Press, 2008); Richard Carr and Bradley W. Hart, *The Global 1920s: Politics, Economics and Society* (London: Routledge, 2016); Julia Stonehouse, *Idols to Incubators: Reproductive Theory Through the Ages* (London: Scarlet Press, 1994); Vern L. Bullough, *Science in the Bedroom: A History of Sex Research* (New York: Basic Books, 1994); Rusterholz, *Women's Medicine*; Jones, *The Business of Birth Control*; Borge, *Protective Practices*; Donna Drucker, *Contraception: A Concise History* (Massachusetts: The MIT Press, 2020).
56. Szreter and Fisher, *Sex Before the Sexual Revolution.*
57. Jon Turney, *Frankenstein's Footsteps: Science, Genetics and Popular* Culture (London: Yale University Press, 1998), 59.
58. Cohen 'Private Lives in Public Spaces', 399–411; Fisher, 'Contrasting Cultures of Contraception', 141–57; Peter Neushul, 'Marie C. Stopes and the Popularization of Birth Control Technology', *Technology and Culture* 39, 2 (1998): 245–72; Claire Debenham, *Marie Stopes' Sexual Revolution and the Birth Control Movement* (UK: Palgrave Macmillan, 2018); Marie Stopes, *Mother England: A Contemporary History, Written by Those who have had No Historian* (London: John Bale, Sons and Daneilsson, LTD, 1929); June Rose, *Marie Stopes and the Sexual Revolution* (London: Faber and Faber, 1992); Marie Stopes, *Dear Dr Stopes: Sex in the 1920s*, ed. Ruth Hall (London: André Deutsch, 1978).
59. 'Family Planning Association', *Lancet* 336 (1990): 171.

60. Not least Helena Wright who worked for the NBCA/FPA for over thirty years and Victor Wallace an Australian obstetrician and gynaecologist; both specifically supported the teaching and provision of contraception. Rusterholz, *Women's Medicine;* University of Melbourne Archives, Wallace Papers, (UOMA/WP).
61. Soloway, *Birth Control and the Population Question*; Soloway, *Demography and Degeneration*; Soloway, 'The "Perfect Contraceptive"', 637–64.
62. Executive Committee Minutes, 29 November 1946. Wellcome Library, Archives of the Family Planning Association (WL/SA/FPA), WL/SA/FPA/A5/4. In the 1950s, the FPA declined an invitate to attend a Swedish-led meeting regarding regular international cooperation on birth control; FPA medical director Helena Wright and her physician son Beric Wright chose to attend and became members of the proposed international committee, later known as the International Planned Parenthood Federation. Leathard, *Fight for Family Planning*, 82.
63. Kathleen Mather, *Hope is Not a Method: A History of the Family Planning Association of Victoria, 1969–1994* (Richmond: McPherson's Printing Group, 1995), 29.
64. NBCA Medical Sub-Committee Minutes, 16 October 1935. WL/SA/FPA/A5/88; Helena Wright, *Birth Control: Advice on Family Spacing and Healthy Sex Life* (London: Cassell's Health Handbooks, 1935), 7; Lucy Bland and Leslie A. Hall, 'Eugenics in Britain: The View from the Metropole', in *The Oxford Handbook of the History of Eugenics*, eds. Alison Bashford and Phillippa Levine (Oxford: Oxford University Press, 2010), 218–9; David Doughan and Peter Gordon, *Dictionary of British Women's Organisations, 1825–1960* (United Kingdom: Routledge, 2013), 48–9.
65. Borell, 'Birth Control Research', 51–2.
66. Sarah S. Richardson, *Sex Itself: The Search for Male and Female in the Human Genome* (Chicago: The University of Chicago Press, 2013), 63–6; Kate Fisher and Jana Funke, 'Sexual Science Beyond the Medical', *The Lancet* 387, 10,021 (2016): 840–1.
67. Iwan Bloch, *The Sexual Life of Our Time in Its Relation to Modern Civilization* (London: Rebman Limited 1908), ix.
68. Carr and Hart, *The Global 1920s*, 59.
69. Neushul, 'Marie C. Stopes and the Popularization of Birth Control Technology', 247.
70. Borell, 'Birth Control Research', 52.
71. Peel 'Contraception and the Medical Profession', 140.
72. C. P. Blacker ed., *International Medical Group for the Investigation of Birth Control, First Report*, 1928. WL/SA/EUG/D/12/15.

73. Matthew Connelly, *Fatal Misconception: The Struggle to Control World Population* (Massachusetts: Belknap Press of Harvard University Press, 2008); Rao, *From Population Control to Reproductive Health*; Ian Robert Dowbiggin, *The Sterilization Movement and Global Fertility in the Twentieth Century* (Oxford: Oxford University Press, 2008); Paul Ehrlich, *The Population Bomb* (New York: Ballantine Books, 1968); Jade S. Sasser, *On Infertile Ground: Population Control and Women's Rights in the Era of Climate Change* (New York: New York University Press, 2018); Thomas M. Shapiro, *Population Control Politics: Women, Sterilization, and Reproductive Choice* (Philadelphia: Temple University Press, 1985); Alison Bashford, *Global Population: History, Geopolitics, and Life on Earth* (New York: Columbia University Press, 2014); Farmer, *By Their Fruits;* Suitters, *Be Brave and Angry*; Foucault, *The History of Sexuality, Vol. 1*; Sophie D. Aberle and George W. Corner, *Twenty-Five Years of Sex Research: History of the National Research Council Committee for Research in Problems of Sex, 1922–1947* (Philadelphia: W.B. Saunders Company, 1953); Richard Cleminson, 'Medical Understanding of the Body: 1750 to the Present', in *The Routledge History of Sex and the Body, 1500 to the Present*, eds. Sarah Toulalan and Kate Fisher (United Kingdom: Routledge, 2013); Ivan Crozier, '(De-) Constructing Sexual Kinds Since 1750', in *The Routledge History of Sex and the Body, 1500 to the Present*, eds. Sarah Toulalan and Kate Fisher (United Kingdom: Routledge, 2013); David M. Friedman, *A Mind of Its Own: A Cultural History of the Penis* (New York: The Free Press, 2001); Renate Hauser, 'Krafft-Ebing's Psychological Understanding of Sexual Behaviour', in *Sexual Knowledge, Sexual Science: The History of Attitudes to Sexuality*, eds. Roy Porter and Mikuláš Teich (Cambridge: Cambridge University Press, 1994); Frank J. Sulloway, *Freud, Biologist of the Mind: Beyond the Psychoanalytic Legend* (Massachusetts: Harvard University Press, 1992); Peter Hegarty, 'Beyond Kinsey: The Committee for Research on Problems of Sex and American Psychology', *History of Psychology* 15, 3 (2012): 197–200; Diana Long Hall and Thomas F. Glick, 'Endocrinology: A Brief Introduction', *Journal of the History of Biology* 9, 2 (1976): 229–33; V. C. Medvei, *The History of Clinical Endocrinology: A Comprehensive Account of Endocrinology from Earliest Times to the Present Day* (United Kingdom: The Parthenon Publishing Group, 1993); Chandak Sengoopta, *The Most Secret Quintessence of Life: Sex, Glands and Hormones, 1850–1950* (Chicago: The University of Chicago Press, 2006); Clarke, *Disciplining Reproduction.*
74. Cleminson, 'Medical Understanding of the Body', 75–7.
75. Cleminson, 'Medical Understanding of the Body', 76.

76. Friedman, *A Cultural History of the Penis*, 258; Hauser, 'Krafft-Ebing's Psychological Understanding of Sexual Behaviour', 210–2; Sulloway, *Freud, Biologist of the Mind*, 277–318; Crozier, '(De-)Constructing Sexual Kinds Since 1750', 143.
77. Hall and Glick, 'Endocrinology: A Brief Introduction', 231.
78. Medvei, *The History of Clinical Endocrinology*, 178.
79. Sengoopta, *Sex, Glands and Hormones, 1850–1950*, 69–70.
80. Clarke, *Disciplining Reproduction*, 123.
81. Margaret Sanger, ed., *Proceedings of the World Population Conference* (London: Edward and Arnold Co., 1927), Announcement.
82. Graeme Gooday, '"Vague and Artificial": The Historically Elusive Distinction Between Pure and Applied Science', *Isis* 103, 3 (2012): 547.
83. Stathis Arapostathis and Graeme Gooday, 'Electrical Technoscience and Physics in Transition, 1880–1920', *Studies in the History and Philosophy of Science* 44 (2013): 203.
84. The International Union for the Investigation of Population Problems: Its Foundation, Work, Statutes and Regulations, 1932. WL/SA/EUG/D/110; International Medical Group for the Investigation of Birth Control: First Issue, 1928. WL/SA/EUG/D/15.
85. Clarke, *Disciplining Reproduction*, 163.
86. International Medical Group for the Investigation of Birth Control: First Issue, 1928. WL/SA/EUG/D/15.
87. Letter O. Mohr to J. Huxley, 12 March 1927. Margaret Sanger Papers, Library of Congress (MSP/LC). Quoted in Borell, 'Birth Control Research', 60.
88. Memoranda Relating to the Proposed Voluntary Parenthood Society, c.1928. WL/SA/EUG/D/202.
89. Edith Howe-Martyn, *Correspondence,* 18 December 1926. Quoted in Borell, 'Birth Control Research', 61.
90. Sanger, *Proceedings World Population Conference*, Announcement.
91. R. Pearl to M. Sanger, 19 April 1926. Quoted in Alison Bashford, 'Nation, Empire, Globe: The Spaces of Population Debate in the Interwar Years', *Comparative Studies in Society and History* 49, 1 (2007): 178–9.
92. Bashford, 'Nation, Empire, Globe', 178 80.
93. W. H. Welch, *Closing Address,* 3 September 1927. Quoted in Sanger, *Proceedings*, 353; Bashford, 'Nation, Empire, Globe', 179.
94. BCIC Memorandum on Work, 1931. WL/SA/FPA/A13/5.
95. BCIC Memorandum on Proposed Re-organisation, [c.1931]. WL/SA/EUG/D/12/12.
96. BCIC Sub-Committee Draft Report, 1928. WL/SA/EUG/D/12/12.
97. BCIC Sub-Committee Draft Report, 1928. WL/SA/EUG/D/12/12.
98. BCIC Draft Statement, [c.1930]. WL/SA/FPA/A13/5.

99. Olszynko-Gryn, 'Technologies of Contraception and Abortion', 541.
100. BCIC Draft Statement, [c.1930]. WL/SA/FPA/A13/5.
101. BCIC Draft Statement, [c.1930]. WL/SA/FPA/A13/5.
102. BCIC Draft Statement, [c.1930]. WL/SA/FPA/A13/5.
103. BCIC Draft Statement of Work, 1930. WL/SA/FPA/A13/5.
104. BCIC Radiological Experiments, General Explanation, [c.1932]. WL/SA/FPA/A13/5.
105. BCIC Present Scope of Work, [c.1929]. WL/SA/FPA/A13/5; BCIC Draft Statement of Work, 1930. WL/SA/FPA/A13/5.
106. Report to the BCIC by H. M. Carleton and H. W. Florey, [c.1930]. WL/SA/FPA/A13/5.
107. BCIC Draft Statement of Work, 1930. WL/SA/FPA/A13/5; BCIC Present Scope of Work, [c.1929]. WL/SA/FPA/A13/5.
108. BCIC Draft Statement of Work, 1930. WL/SA/FPA/A13/5.
109. Report of Berlin Visit Investigating the Gräfenberg Ring Contraceptive, December 1929. WL/SA/FPA/A13/5.
110. Rusterholz, *Women's Medicine*, 4.
111. Graeme Gooday, '"Nature" in the Laboratory: Domestication and Discipline with the Microscope in Victorian Life Sciences', *The British Journal for the History of Science* 24, 3 (1991): 309.
112. BCIC Draft Statement of Work, 1930. WL/SA/FPA/A13/5.
113. Rusterholz, *Women's Medicine*, 68.
114. BCIC Memorandum on Proposed Re-organisation, [c.1931]. WL/SA/EUG/D/12/12.
115. Memorandum on the Suggested Framing of Rules for the SPBCC, 1926. Wellcome Library, Archives of the Private Papers of Margery Spring Rice (WL/SA/SR), WL/SA/SR17/1–11.
116. Leathard, *Fight for Family Planning*, 48–9; Debenham, *Birth Control and the Rights of Women,* 159–60.
117. BCIC Memorandum on Proposed Re-organisation, [c.1931]. WL/SA/EUG/D/12/12.
118. Memorandum on the Suggested Framing of Rules for the SPBCC, 1926. WL/SA/SR17/1–11.
119. NBCC Amended Draft of Constitution, 1930. WL/SA/EUG/D/12/12.
120. That is not to say that aspects of eugenic instruction and theory didn't reach individual clinics at one time or another, but it never formed any significant part of the association's prescribed and standardised clinic structures, functions or services. The North Kensington Clinic did flirt with Eugenic Sessions in early 1960 (WL/SA/FPA/NK/32) however by this time the meaning of eugenic was somewhat changed and service was exclusively offered to self-referrals who worried about the possible transmission of personal or familial mental or physical weaknesses to offspring, or those specifically referred by a physician.

121. Memorandum and Articles of Association of the Eugenics Society, 1926. Wellcome Library, Archives of the Private Papers of Carlos Paton Blacker (WL/PPCPB), WL/PPCPB/K/1.
122. Memorandum of Grants Recommended by the General Purposes Committee, 6 March 1934. WL/PPCPB/J/4/1.
123. *Family Planning and Family Planning Clinics Today*, May 1962. WL/SA/EUG/P/39/42.
124. *Handbook for Speakers Issued by the Joint Committee on Voluntary Sterilization*, [c.1934]. WL/PPCPB/B/5/3; Eugenics Society, *Four Generations of Insanity and Mental Deficiency*, [c.1935]. WL/SA/EUG/G/30/15/7; Eugenics Society, *Pedigree of a Woman who Married Twice*, [c.1935]. WL/SA/EUG/G/30/32/3; Eugenics Society, *Environment versus Heredity*, [c.1935]. WL/SA/EUG/G/33/5; Eugenics Society, *Marriage of Normal with Feebleminded Persons,* [c.1935]. WL/SA/EUG/G/30/35/10; Eugenics Society, *Environment and Heredity,* [c.1935]. WL/SA/EUG/G/30/35/4.
125. *Handbook for Speakers Issued by the Joint Committee on Voluntary Sterilization*, [c.1934]. WL/PPCPB/B/5/3; Eugenics Society, *Four Generations of Insanity and Mental Deficiency*, [c.1935]. WL/SA/EUG/G/30/15/7; Eugenics Society, *Pedigree of a Woman who Married Twice*, [c.1935]. WL/SA/EUG/G/30/32/3; Eugenics Society, *Environment versus Heredity*, [c.1935]. WL/SA/EUG/G/33/5; Eugenics Society, *Marriage of Normal with Feebleminded Persons,* [c.1935]. WL/SA/EUG/G/30/35/10; Eugenics Society, *Environment and Heredity,* [c.1935]. WL/SA/EUG/G/30/35/4.
126. Olszynko-Gryn, 'Laparoscopy as a Technology of Population Control', 147–8.
127. Leathard, *Fight for Family Planning*, 59.
128. Representatives of the Governing Body of the NBCA and the Council of the Eugenics Society Meeting Minutes, 3 December 1936. WL/SA/EUG/D/12/16/23.
129. The Eugenics Society and National Birth Control Association, 1936. WL/SA/EUG/D/12/16/23; Recommendations as to the Future Relation of the NBCA and Eugenics Society, April 1937. WL/SA/EUG/D/12/16/19.
130. Letter L. Darwin to C. P. Blacker, 28 April 1937. WL/SA/EUG/D/12/16/23.
131. NBCA and Eugenics Society Meeting Minutes, 8 July 1937. WL/SA/EUG/D/12/16/23.
132. Letter M. Pyke to C. P. Blacker, 9 November 1937. WL/SA/EUG/D/12/16/23.

133. J. P. Stocks, 'A Conference on Birth Control', *Nursing Times* (1934). WL/SA/EUG/D/12/16/20.
134. Draft Memorandum to Ministry of Health, February 1937. WL/SA/EUG/D/12/16/19. Their underline.
135. Ministry of Health, *Circular 1622 Maternal Mortality* (London: Ministry of Health, 1937).

CHAPTER 2

Instituting and Regulating the Contraceptive Clinic and its Services

Abstract This chapter will discuss the origins of the North Kensigton Women's Welfare Centre as a pioneering site of medical, scientific and technological research and development. Through this site, the National Birth Control Association/Family Planning Association sought to align its medicalised contraceptive clinic and administrative practices with contemporary general medical protocols and the services offered by Ministry of Health-directed Local Health Authorities. It further aimed to develop and export its medical and scientific contraceptive programs, and standards for clinic arrangement, function and prescription within the increasingly lenient parameters the Ministry allowed.

Contraceptive historians have spent decades investigating and writing about birth control clinics and their impact on society, demographics, sex, politics, women's health and wellbeing, and culture.[1] The sites are occasionally described with some detail, but for the most part, there is an assumption that readers already know what they looked like, why and how they were established in specific localities, and their structure and function. However, surprisingly little focus has been paid to clinics as functioning sites of medico-scientific investigation and education, let alone as standardised therapeutic facilities operating within the accepted bounds of extant British medical structures. From very early

N. Szuhan, *The Family Planning Association and Contraceptive Science and Technology in Mid-Twentieth-Century Britain*, Medicine and Biomedical Sciences in Modern History, https://doi.org/10.1007/978-3-030-81300-0_2

in its history the National Birth Control Association (NBCA) was keen to standardise its contraceptive clinics and develop systems and models that could be implemented in all affiliated sites and exported widely. Through a case study of the North Kensington Women's Welfare Centre (NKWWC), the largest NBCA affiliated clinic, this chapter investigates how the association worked to have its clinics viewed as legitimate sites and sources of scientifically sound medical and technical investigation, regulation, therapy and education, and to impact how contraceptive technologies were understood, prescribed and used therein.

During its first decade, the North Kensington clinic became the association's primary site of applied scientific and clinical investigation into women's sexual, maternal and contraceptive health. Using statistics, services and products at the site, clinicians and scientists demonstrated how contraceptive technologies could be employed as a preventive therapy with mass and diverse public health benefits. This medicalisation profoundly changed contraceptive provision. Once medical and scientific standards became the accepted means through which contraception was to be understood and prescribed, a variety of programs and policies emerged from the North Kensington centre that allied the practice to public health and ultimately forced the Ministry of Health to officially recognise contraception as a legitimate and necessary therapy.

Identifying and Exploiting a Medico-Scientific Niche for Contraceptive Clinics

The NKWWC located at 12 Telford Road was founded on 6 November 1924 by political and social reformers affiliated with the Society for the Provision of Birth Control Clinics (SPBCC).[2] The endeavour was led by Margery Spring Rice a renowned contraceptive advocate and social activist who believed access to safe and effective birth control would greatly improve women's collective health. This site would address the 'appalling levels of poverty and overcrowding in North Kensington',[3] a formerly 'fashionable residential area', that had by the 1920s become a slum district, with rows of apartments, many sharing communal facilities.[4]

The clinic was established on the 'evacuated premises' of the Mary Middleton and Margaret MacDonald Baby Clinic which provided infants and children diverse health services 'without any charge … of any kind'.[5] Amenities included medical assessment and intervention, food medicines, dental and minor surgery, and a 'dispensing druggist'.[6] Importantly, the centre also offered pre- and antenatal medical advice to mothers.[7] This was no makeshift clinic, or crèche, but a functioning medical facility. In 1912, within a year of opening, the baby clinic boasted two regular doctors, Annie L. Kann and Ethel Bentham, and prided itself on providing both a breadth and level of continuity of care that most contemporary hospitals could not. The well-established and equipped centre was abandoned solely due to demand outstripping capacity for service at the Telford Road location. As a result, the baby clinic relocated in July 1924 to a 'new and more commodious premises'.[8] Therefore, when the SPBCC was looking for a site on which to establish its contraceptive clinic, this former Local Health Authority run site proved 'fairly well equipped' to facilitate the founder's aims 'to give scientific birth control advice to women who could not afford the fees of specialists'.[9]

Rice superintended the NKWWC between 1924 and 1958. She structured its evolution, and especially encouraged a female-centred and directed medical focus.[10] She and the management esteemed the medicalised approach Marie Stopes employed in her Mothers' Clinic for Constructive Birth Control. The management was particularly keen to establish a site that, like Stopes', facilitated women's ability to choose whether and when they became mothers by instructing them in 'how to regulate [their] own fertility'[11] and offering maternal and infant care and contraceptive services in a medical environment.[12] The management, and indeed the emerging women-led contraceptive movement, endeavoured to model and project a female self-empowerment focus that would provide women with basic intimate knowledge of their bodies, sex and fertility.[13]

The advent of the Maternal and Child Welfare Act of 1918 generated large-scale official and voluntary efforts to acknowledge and attend the health and basic needs of mothers and their offspring. Lara Marks and Dorothy Porter have demonstrated that mother and infant welfare was of increasing importance to the British state during the interwar period as the relationship between welfare and fertility, and social, political and medical issues became better understood and publicised.[14] As women's prime importance to the state was in their capacity as mothers—fertility, pregnancy and birth had to be made as safe as possible; and so birth control and spacing became increasingly linked to the concept of maternal health. From 1925 contraceptive campaigners began to tap into concerns about maternal health and population growth and quality, by highlighting the fact that some women needed to prevent or postpone childbearing and that sites like the NKWWC offered safe and effective methods of doing so.

Before 1930, one significant problem for contraceptive advocates was the government's unwillingness to allow local health authorities to provide birth control advice, let alone instruction and products. So, from the late 1920s campaigners agitated for government support for medically necessary contraceptive assistance.[15] This was effected through the collective activism of 'large bodies of organised women', other voluntary societies and occasionally municipal authorities; they demanded contraception be made available to all mothers through Ministry of Health welfare centres as a preventive health tool for women. This focus on prevention was a significant and increasing focus of modern public health theories and practices from the mid-nineteenth century onwards. During the subsequent century it found its zenith in the fusion of principles of fiscal and eugenic fitness that promoted the idea that the national public and economic health were both better served by preventing the introduction of contraband that might diminish the vitality of either for the greatest collective good.[16] It was this increased acceptance of preventive care and active incorporation of women citizens into the purview of national public health that paved the way for the development of an increasingly comprehensive medical service at the North Kensington site.

Shortly following the establishment of the NKWWC, the Kensington Council instituted one of the most liberal approaches to contraception in Britain.[17] In 1926, at Rice's urging, it unsuccessfully petitioned the government to allow its infant and maternity clinics to incorporate contraceptive instruction. It claimed that relaxed opposition would 'guarantee greater continuity of care' by consolidating all maternal health services.[18] This agitation and sympathy peaked in April 1930 at the Westminster Conference on the Giving of Birth Control Information by Public Health Authorities. There were delegates from 172 welfare organisations including maternity and child welfare centres, contraception clinics and other national organisations, who joined public health representatives from Kensington to formally challenge this position.[19] The group expressed particular support for contraception to be provided to women whose health would be compromised by future pregnancy. This groundswell of semi-official pressure helped the NKWWC and National Birth Control Council to identify and formally pivot their agitation and services to address a pressing need for scientifically sound birth control education and methods.

Whilst existing contraceptive providers worked to hone the products and services offered in clinics, the agitation of public health and welfare groups pressured the Ministry of Health to take a stand on the issue in 1930. In July it published Memorandum 153/MCW which defined the powers and limitations of local health authorities to provide contraceptive guidance and products. The memo asserted it 'was not the function' of government-funded welfare clinics to 'giv[e] advice to women on contraceptive methods',[20] but it included an important caveat that would prove vitally important to the NKWWC and its parent organisation. The memo permitted local health authorities to provide contraceptive advice

> (1) only for women who are in need of medical advice and treatment for gynaecological conditions; and (2) ... only to married women ... in whose case pregnancy would be detrimental to health.[21]

This vaguely permissive edict officially provided conditions for the public to access contraception, provided local health providers would acquiesce.[22] However, initially very 'few local authorities ... troubled themselves to use even the facilities they [were] allowed!'.[23] Some medical officers declared the Memo confusing, others lacked funds, and yet more claimed to be 'keenly sensitive to the religious and moral sensibilities of their constituents' whose opposition was acute.[24] The result was that after fifteen months only 'thirty five [of the existing 1200] local authorities ha[d] authorised the giving of advice on contraceptive methods'.[25]

This ambiguous Memo forced the Ministry to issue two further statements clarifying its position. The first in July 1931, Circular 1208, clarified that decisions regarding contraceptive provision were 'entirely within [local authority] discretion'. But also stipulated that the Minister had determined this work 'should not be regarded as falling with the scope of [medical officer's] normal duties', and thus individuals were 'free to undertake or decline it'.[26] The Circular reiterated that only serious gynaecological risks necessitated preventive contraceptive advice.

In 1928 a Departmental Committee of Maternal Mortality and Morbidity was appointed by then Minister of Health Neville Chamberlain, to undertake a sociological and medical study into the stagnation of maternal mortality rates. This study presented an opportunity to stress the links between too often and closely spaced pregnancy and the increased risk of death and disease. In June 1930 an interim report was issued supporting the need to circumvent pregnancy as the 'primary avoidable factor' in the category 'deaths not primarily due to pregnancy'.[27] Two years later the committee issued a final report that had scrutinised the causes of 3085 maternal deaths. The investigation determined that 13.5 per cent or 514 actual deaths, directly resulted from 'diseases definitely made worse by pregnancy'. The committee's finding directly challenged the Ministry's gynaecology-based limitation on the provision of contraceptive products and advice. It also demanded that 'special attention [be given to] the avoidance of pregnancy by women suffering from [various] organic diseases' and in accordance with Circular 1208

> that advice and instruction in contraceptive methods should be readily available for such women, and their husbands, from private practitioners at hospitals, or at gynaecology clinics set up by Local Authorities.[28]

This determination compelled the Ministry to issue a third and final clarification in 1934, Circular 1408, which pointedly increased the conditions under which health authorities could offer contraception to include 'other forms of sickness, physical and mental'. Important for the medical legitimisation interests and efforts of the NBCA and NKWWC, the Circular concluded with the unambiguous assertion: 'what is, or is not, medically detrimental to health must be decided by the professional judgement of the registered medical practitioner in charge of the clinic'.[29] This Circular undeniably united contraception with female health in public health legislation and parlance and ascribed medical and social value to women not directly tied to childbearing. It formally afforded doctors the discretion to freely interpret 'detrimental to health' and 'likely [to] seriously endanger life' to evaluate on merit every patient's contraceptive need. But, this success did not transform the medical profession's overwhelming opposition to and disconnection from contraception. Rather the Ministry's encouraging position marked a decisive but minute success for the NBCA and its clinics. It provided a framework for the association to undertake and advertise contraceptive treatment, prescription and research under a Ministry-sanctioned medical banner. This pivot towards women's health ultimately resulted in contraception becoming comprehensively integrated with medical science. The association's next task was to liberally provide this medical service to the British public from its model clinic: the NKWWC.

The North Kensington Women's Welfare Centre's Medico-Scientific Direction

At the inauguration, the clinic was envisioned as an 'independent, co-ordinating body to advance the movement and to provide support for clinics to promote [contraceptive] medical research'.[30] Achieving this goal was difficult, as many physicians and scientists refused to add legitimacy to contraception by engaging with untested and unverified contraceptive technologies. Thus, ongoing opposition to contraception as a health practice prompted then disparate and localised British birth control providers to unite in an effort to provide means, locations and funds to expand pure and applied scientific and medical research. Owing to the foresight of its political and medical staff, by 1930 the North Kensington clinic was already a flourishing site of medical provision and investigation.

The archival history of the NKWWC's first decade is sparse. There is little record of the site's management during its opening and the foundation and consolidation of its medical committee and research programs. Lara Marks and Audrey Leathard both characterise the site as a flourishing hub of medical provision and scientific research during its first decade. Statistics are limited prior to the mid-1930s, and the remaining records before 1934 only feature total clinic attendees.[31] That decade, annual attendance increased by approximately 100 women, beginning with 255 attendees in its initial nine months of operation and peaking at 1082 in 1933–1934. In 1929 Norman and Vera Himes explained that clinic attendance got off to a slow start owing to an 'absence of extensive propaganda' and that this meant that many local residents' were unaware that a clinic had even opened. Further, many that knew of the site 'were much too shy' to attend as they were uncertain about clinic appointments, 'but gradually more and more came… [and] they found that after all it was not so alarming'.[32] Despite this, enough women attended to mark the venture a success and by 1933, the centre had officially become the most comprehensive in Britain. It operated five days a week, and offered four ninety-minute birth control, and three ninety-minute gynaecological sessions[33]—then a trial service that would be formalised the following year.[34] As the NKWWC was almost fully functional upon foundation, the consolidation and expansion of its contraceptive and then women's medical programs began directly. To supplement the work of its executive committee that managed the clinic's function, affiliation and politics, a local medical committee was entrenched by 1934. That August it held its 'Seventeenth Meeting',[35] that proved it was already deeply engaged in linking the centre's contraceptive activities directly to matters of women's health. This committee guaranteed that standardised contraceptive services and medical therapies were effected at the clinic, and scientifically sound, patient-appropriate educational programs were offered, and later founded marriage and sex counselling and sub-fertility clinics. Finally, the group established standards for doctor training, record keeping, clinic management and structure, and patient contact, to assess and ensure contraceptive acceptability and effectiveness.

The committee was founded in 1927 under the chairmanship of Helena Wright, the centre's newly appointed chief medical officer of health.[36] The medical committee records prove that many contraceptive and women's health initiatives emanated from the group. These were: the compulsory gynaecological examination program which preceded contraceptive prescription; clinical trials and the maintenance and reporting on clinical statistics and data; medical treatments and referrals; standardised completion and maintenance of patient records; medical and lay teaching programs; and quality and safety tests and controls for specific contraceptives. Each became key NKWWC and NBCA/Family Planning Association (FPA) programs between 1930 and 1955.

This committee positioned the NKWWC at the vanguard of the NBCA's self-regulatory ideals through a series of programs that were designed to medicalise and guarantee the technologies and techniques recommended by the clinic. Towards that end it appointed Scottish scientist Cecil I. B. Voge to test the spectrum of contraceptive products and devices it prescribed. Voge was currently researching spermicides for an efficacy study jointly funded by the American National Committee on Maternal Health and Bureau of Social Hygiene. The data he produced at North Kensington would then inform and shape the products and methods it would distribute and its clinical trial and evidence-based advertising and medical teaching agendas. These research programs, which actively unified the British and American contraceptive agendas, were intended to 'mark the beginning of a program for the control by the medical profession of contraception'.[37] The scientific focus on each side of the Atlantic diverged significantly after 1935. Whilst the Americans worked to develop contraceptives that demanded medico-scientific oversight and control, the standardisation and perfection of current contraceptive technologies remained the focal point of the NBCA and its British efforts.[38] The association's work to this end is detailed in chapters 4 and 5.

Superintendent Rice openly dedicated the North Kensington centre to facilitating and actively applying contraceptive research in order to report to the public and medical profession that the modern technologies of birth control were verified.[39] She and the executive committee understood the clinic had a unique capacity to undertake laboratory and statistical studies that could improve its products and standards, and so requested the medical committee define exactly how this could be achieved.

i. To what extent can this be a teaching centre?
ii. What backing do the medical officers need in the way of laboratory workers, etc., to maintain this object?
iii. The Executive Committee realises the need of up-to-date knowledge, etc., to what extent do the B.C.I.C. [Birth Control Investigation Committee] fulfil this need?
iv. How far can the medical officers cooperate in the reorganisation of the clinic, with particular reference to the teaching side?[40]

This bold scientific and regulatory agenda was realised through the institution and expansion of research and training programs at the NKWWC after 1930. As the North Kensington site then operated in a grey area, somewhere between an independent clinic and a full NBCA affiliate, its operations and activities were similarly composite. Its goals and tactics were informed by the Birth Control Investigation Committee (BCIC), and its successes were subsequently adopted as NBCA core objectives. The clinic hence became the form and management prototype for association clinics.

From October 1934, Voge became the medical committee's scientific point man. He pioneered and undertook to test and report spermicidal and later rubber contraceptives for harm and efficacy. His findings were collated into data relevant to the broader medical and scientific community and were intended to stimulate the British Royal Society of Medicine and newly formed Royal College of Obstetricians and Gynaecologists to 'develop… a plan for the control of the contraceptive industry in the interest of a sound practice of preventive medicine'.[41] But despite the enthusiastic acceptance of Voge's work by its American funders and elaborate visions for its US application, the President of the latter, Eardley Holland asserted the 'Society... were not interested in the det ails (sic) of contraception until some perfection of method had been found'.[42] However, he was interested in 'kindred subjects', and invited Voge to lecture on 'the PH of the Vagina, with special reference to fertility and sterility both natural and artificial'. Holland also agreed that any ensuing discussion of contraception 'would be allowed to proceed', and so the NKWWWC medical committee sent its medical staff to hijack the post-lecture discussion to steer the conversation to the topic. Armed with Voge's scientific evidence in favour of the practice, the committee admitted that although it may be 'worthwhile to force ... contraception on the medical profession as a whole, general co-operation was wanted rather than isolated efforts'.[43] In a decade where Claire Jones and Jessica Borge have demonstrated the contraceptive landscape and market was rapidly developing, the clinic and association medical staff united to announce Voge's scientific results and inform doctors of 'reliable measures' they could confidently prescribe.[44]

Despite Voge's findings, the medical profession remained broadly unconvinced of the medical nature of contraception. In response, he expanded his North Kensington work to build a market for his findings regarding proprietary contraceptives through a web of cooperative manufacturing chemists, medical journals, interested physicians and birth control advocates.[45] Correspondence and records demonstrate that Voge zealously networked to advertise the NKWWC and its suppliers and methods, as well as his findings within medical and scientific circles. His primary achievement in this regard was convincing the *British Medical Journal*, *Lancet* and *The Practitioner* to include contraceptive advertising only 'if their presumptive harmlessness and relative efficacy had been assured, and the truthfulness of the advertising accredited'.[46] As Jones has shown, during the 1920s a comprehensive industry-wide shift toward advertising contraceptives in the medical presses occurred. This would, of course, more formally ally the practice and its technologies with medicine and concurrently seemed to signal a professional openness to there being a role for medicine in birth control.[47] In further support of this idea, Voge read the NKWWC medical committee 'letters from various journals' indicating they 'were relying on him' to certify each advert and product received. This was clearly a blatant attempt to elevate the importance of his own expertise, research agenda and necessity to the centre and broader movement, but also prompted the committee to agree that publicising any 'definitely negative' results of Voge's studies was in medical and public interest. The committee advised Voge to notify the journals that he had determined that 'all foaming tablets were unreliable and undesirable'.[48] All agreed that even if this action 'would not be very far-reaching in effect, [it] would be taking a definitive stand as a body' to support a 'definitive drive toward [contraceptive] standardisation'.[49] This is the first recorded example of the NBCA, or its affiliates, citing officially commissioned scientific research to approve or denounce specific contraceptives.[50] Finally, Voge liaised between his patrons in the United States and the United Kingdom and brokered an agreement to share the results of his research to the benefit of each. This trans-Atlantic collaboration was a boon for each party, it unified the pure and applied scientific studies being funded and spearheaded by the primary contraceptive advocates in the United States and the United Kingdom, and enabled Voge to promote his testing methods widely.[51]

Voge's inclusion on the committee galvanised the group and encouraged it in its belief that scientific and medical cooperation was the future of contraception. His star was ascendant and his reach and influence perceived to be so wide that the NBCA Medical Sub-Committee questioned 'the extent to which Voge [would be made] available', as his North Kensington work 'was being withheld from [its] committee'.[52] This interaction reveals the tenuous strength of the alliance that loosely unified five discrete societies under one title. Although all twenty-eight clinics that operated under the NBCA banner by 1935 were technically association clinics, they did not function as 'a courageous, brilliant band', but rather worked towards similar goals without uniformity of form or function.[53] A definite problem for a group attempting to shape and direct the national response to and provision of birth control. In 1935 the Society for the Provision of Birth Control Clinics that founded and managed the North Kensington site had still not fully amalgamated into the NBCA; it was represented on the association's Governing Body and Executive Committee and was working to cultivate a 'very close' working relationship.[54] This meant that despite the NBCA's request for access to Voge, there was no obligation for the NKWWC to make him or his work directly accessible—even if they could. As it happened, the NKWWC was not at liberty to autonomously navigate this relationship. In juggling distinct transnational patrons and allegiances, Voge had negotiated for the NKWWC to use his American sponsored research, but that 'any other use

… must be first referred to Voge to permit or not at his discretion'.[55] Voge attempted to cultivate an international reputation for himself as a dispassionate expert on contraceptive technologies and practices, and establish a distinct medico-scientific niche for his work. This would mark him as the preeminent contraceptive scientist in the Western world, a position that for him promised both professional and financial gain and heightened his appeal to the NBCA as an asset in its goal to fuse contraception and science, and the NKWWC as its new scientific celebrity.[56] Once the North Kensington committee put the request to him, Voge graciously responded that 'he was entirely agreeable with any reasonable circulation of his findings' as long as credit went to the NKWWC and New York-based Committee for Maternal Health.[57]

Voge was unquestionably interested in cultivating a strong relationship with the association. In 1935 he began to stress a political outlook that the 'work [he and his patrons] were doing on this matter [using science to legitimise and standardise contraceptives] should be carried out under the auspices of an authoritative body'.[58] Voge recognised the NBCA as the ideal arbiter of the British contraceptive market; due to it fast becoming the primary national distributer of contraceptives and work developing a scientific niche for contraception. In pursuit of this agenda and to pressure the association towards decisive regulatory action, Voge allowed the NKWWC to furnish the NBCA with his findings, as well as correspondence from manufacturing chemists and journal articles 'referring to various tests of reliability and suitability of certain contraceptives' that he had, until now, addressed alone.[59] Ultimately, the association was persuaded to attempt to standardise and oversee the local contraceptive market, but employed its own researcher and methodology.

Voge remained the NKWWC scientist and began to expand his spermicidal work to investigate specific proprietary chemical contraceptives and also test rubber devices, specifically cervical caps, their quality, longevity, spring tensions and dome heights.[60] This line of enquiry emerged in response to concerns regarding 'the quality of the Dutch pessaries supplied by Messers Lamberts Prorace'. This was a particularly disturbing exposure as Lamberts, one of the rubber contraceptive manufacturers established in East London, was the NKWWC's preferred cap supplier.[61] Voge examined Lambert's new 'Double Rubber' latex cap which was re-developed to 'reinforce the pessaries by using a double thickness of rubber for the dome' to prevent failures and guarantee the durable quality of the product.[62] His testing mandate was to examine 'what effect [was] made on [the cap] by ... vaginal secretions [and] ointment and soluables' as some natural and unnatural substances had been shown to degrade rubber. Voge reported almost immediately that the product proved 'undesirable', and the study was extended to other brands.[63]

Voge cultivated a close working relationship with another local manufacturer named Gilmont Products, and brokered a direct collaboration with NKWWC physicians to design and develop 'a really satisfactory cap'. After testing Gilmont's caps against Lambert's, Voge declared the new manufacturers existing range of caps adequate but stressed they could be improved. As the local medical committee was keen to elevate standards for contraceptives it prescribed, Gilmont's interest in developing a contraceptive with the clinic medical staff was probed. The manufacturer confirmed its 'willingness to try to fulfil the doctor's wishes' regarding the specific caps desired, 'so long as these [wishes] were made perfectly plain'.[64] In a quid pro quo arrangement, Gilmont agreed to develop a contraceptive to meet North Kensington's requirements in exchange for a promise that its caps would become the site's preferred contraceptive. This relationship constituted an early form of corporate sponsorship that was intended to raise Gilmont's brand popularity through a combination of expert and user endorsement and NKWWC-led market saturation.[65] With this understanding in place, Voge and the committee determined which features the ideal cap should boast. Gilmont agreed to produce uniform caps in both 'soft and reasonably hard' varieties, that featured a 'high dome... [that was] roughly 2/3 of the [rim] radius' and a firm and steady spring tension. The Committee was galvanised by the prospect that in collaboration with a dedicated clinic-directed manufacturer a 'perfect cap might be ultimately produced'.[66]

After four months of research and development, Voge announced that although Lamberts caps were intermittently superior to Gilmont's, its products were not uniform and therefore his tests could not guarantee the quality or efficacy of those caps with scientific certainly. This finding bolstered the perception there was an important and pressing problem that Voge, the NKWWC and Gilmont were addressing through the collaboration—a threat to their collective contraceptive ambitions presented by lax manufacturing and efficacy standards. As Jones has shown, this perception certainly existed, but estimates of failure rates varied so greatly that only the principle that existing products were inefficient could be agreed.[67] This was a severe blow for Lambert's; as owing to the NKWWC committee's resolution that product standardisation was vital to its prescription and renewal requirements, the brand's products were judged unsuitable for use.[68] In November 1935 Voge confirmed that Gilmont's new cap had met all the NKWWC's specifications, at one-third of the cost of Lambert's products. The committee approved and formalised Gilmont Caps as the site's preferred barrier contraceptive.[69] Some historians, notably Borell, Löwy, Borge and Jones, have documented scientific and professional competition in the contraceptive market, typically as it relates to sales; but as yet, no other detailed study of one manufacturer successfully ousting another from any market exists.[70] That omission from the historical record marks this murky alliance between Voge, Gilmont and the NKWWC as a remarkable example of the emerging symbiotic relationships between contraceptive business and medicine, and (supposedly) disinterested scientific study and industry in the interwar period.

During the mid-1930s Voge ingratiated himself at the NKWWC and the medical committee relied heavily on his research and counsel to structure and design clinical and regulatory techniques. The North Kensington physicians especially valued having immediate access to a laboratory scientist, and many pure and applied research projects were founded on the understanding that Voge's affiliation would be ongoing. However, in 1936 Voge abruptly ended all contact with the site and indeed the NBCA, without renouncing his research or medical committee position. That year, Voge's dedication to contraceptive standardisation declined without explanation. Between February and September his manufacturer reports were continually deferred, and the results of his effectiveness tests of Prensol and Clinicol Soluables were not submitted until September. In November, Voge was absent without apology and ignored the committee's request for the details of his techniques and 'sperms used to compare with other work done', presumably then being spearheaded by the NBCA and BCIC.[71] In April 1937, the medical committee, upon learning he had taken a role in private industry, accepted that Voge had 'severed his connection with the Birth Control Clinics'. Not letting this setback impact the very important medico-scientific work that the site had pioneered, the committee immediately determined another 'part time laboratory worker should be appointed'.[72] The value of directing and implementing laboratory research from a functioning contraceptive clinic was thus demonstrated, and the method was thereafter adopted and extended by the NBCA to inform and improve medical and contraceptive successes.

Within its first decade, the NKWWC benefitted greatly from the availability of a functioning clinic in an appropriate and populous district as well as the appointment of a talented and passionate medical committee and (mostly) dedicated scientist who together foresaw the value of establishing and overseeing contraceptive standards. These factors combined to create a clinic well placed to exploit its medical and scientific programs in furtherance of the legitimisation of contraception. The zeal for scientific standardisation and regulation that resulted from these early clinic developments, was also visible in the development of the site and the programs that flourished therein.

Configuring the North Kensington Clinic and Service in Response to Official Resistance

During the late 1920s, as the North Kensington management and executive staff worked to establish and advertise the clinic as a public health hub with bona fide medico-scientific credentials, the medical profession's apathy towards contraception peaked. Although sympathy and interest was observed amongst an emerging generation of doctors, a broad and definitely generational hostility remained towards arguments that physicians were the 'natural guardians of family health' and should lead the crusade for contraception.[73] The entrenched physicians generally and honestly cited inadequate knowledge and a lack of experience with the modern 'scientific' technologies. These older physicians often also irresponsibly compounded their avowed inadequacies by refusing to direct eager patients to contraceptive clinics like the NKWWC that had emerged to fill the void. This was still the case in 1939 when North Kensington superintendent Rice offered a scathing assessment of the medical profession's progress regarding contraception for the Women's Health Enquiry Committee. She predicted a burgeoning public health catastrophe that was caused by 'deplorable ignorance or prejudice on the part of the professional medical attendant'.[74] Through her work with NBCA and NKWWC medical committees and directors, Rice attempted to remedy medical ignorance by incorporating a series of 'lectures to doctors' and pressuring the profession to accept the practice by aligning the NKWWC's form, function and services with modern medical sites and practices. This action had two clear benefits, it created a reproducible model site to trial and refine NBCA medico-scientific programs and linked clinic services with general medical practice for patients.

In 1926 the medical profession's objections to contraception were pronounced in *Medical Views on Birth Control*. Eight practising physicians contributed to the tome in a far-reaching appraisal of the social, political and medical aspects of contraception, that ultimately downplayed its potential benefits, and stressed the 'endless… possibilities of physical harm' believed to be associated with contraceptive technologies and practices.[75] The only optimistic assessment was voiced in the introduction penned by renowned physician Sir Thomas Horder, King Edward VII's personal doctor who also taught and treated many of Britain's most prominent clinicians. He later became the first NBCA/FPA President and led government and voluntary social welfare and health committees including the Family Relations Group and the Eugenics Society.[76] Horder explained the practice offered a means to assuage human misery and improve general health, and appreciated that the publication, despite its overall gloomy tone, represented a new 'spirit of scientific enquiry' for physicians that had yet contributed little to contraceptive knowledge and literature. Horder's optimism was countered by the medical contributions that articulated little 'personal conviction [and] … still less inclination' towards contraception. Some offered tentative approval where it intersected with medical need, whilst others 'exhibit[ed] quite frankly an element of revulsion'.[77] Consulting gynaecologist Dame Mary Scharlieb, the only female contributor to the tome, claimed

> the whole aspect and story of artificial control appears to be sordid and unnatural, and when the immediate risks and the probable future consequences are realised it is difficult to understand that anyone should be found willing to practise or advise such methods.[78]

Her article explained that 'the undesirability of conception control is clearly proved from the point of view of health'. Scharlieb stressed alleged risks inherent to the use of specific contraceptive technologies such as the intra-uterine pessary. Highlighting its 'murderous' and 'abortifacient' effects she dubbed it a 'very dangerous instrument' with serious potential side effects that included sepsis, endometriosis, peritonitis and occasionally even death. In her zeal to comprehensively condemn the practice, Scharleib cynically focused her essay on the only contraceptive device then being tested and prescribed that blurred the line between legal contraception and illegal induced miscarriage or abortion.[79] Her's was just the most egregious of a litany of accusations presented by the various contributors that alleged that 'nervous' as well as 'real physical' injuries (even death) might result if couples employed barrier contraceptives, or if 'instruments, whether vulcanite, metal or glass' were inserted into the vagina or uterus of an 'inexperienced woman' and were accidentally left 'in situ for days, [or] even weeks'. Finally, a psychological impact was hypothesised from the lack of romantic spontaneity compelled by the necessity to prepare for conception-safe copulation, rendering the activity more 'chemical experiment' than the expression of the 'purest, highest love'.[80]

The older generation of physicians, that then represented the whole, who held such views contended that it was not the profession's duty to engage with risky practices and technologies that might weaken marital, sexual and family bonds. This view allowed some more senior doctors to evade any professional obligation regarding contraception, except in cases where a genuine risk to health existed. However, in 1930 at the British Medical Association's Annual Representative Meeting official leeway was granted to doctors willing to engage with contraceptives. It determined that

> every ... medical practitioner, has the right to advise either for or against the use of contraceptive methods in accordance with individual judgement and responsibility and should not be subject to dictation ... in this matter.[81]

Like the Ministry of Health's contemporary memos and circulars, this ruling appeared to provide carte blanche to medical professionals ministering at risk gynaecology patients to prescribe contraception to prevent future pregnancy-related illness. But that was not how the decree was interpreted. In *The Principles of Contraception* (1935), Joan Malleson, an NKWWC physician and sexual health and fertility specialist, explained that many North Kensington patients reported that doctors had warned them that further pregnancy could be life threatening but 'never told [them] what to do' to prevent it.[82] This response did not represent a failure of the physician in his duty to protect patient health. These medical attendants had 'represent[ed] to the patient the need for the avoidance of the pregnant state', but citing the 'absence of adequate analysis' regarding the safety and efficacy of current contraceptive methods could or would not offer any prescription. In the absence of definitive knowledge about the relationship between contraception and risk, doctors relied on personal and published clinical experiences that seemed to confirm a correlation between contraceptive users and 'inflammatory diseases of indeterminate aetiology'. Thus, the public figures of the profession adopted a distinctly oppositional posture towards contraception, that led Malleson to claim that 'the relationship of the medical profession to the public [in this regard was] very unsatisfactory'.[83]

Contemporaneously, the NKWWC medical director Helena Wright issued *Birth Control* (1935), a plea to the public to employ available contraceptive services and demand state and medical responsibility and oversight.[84] The monograph correctly identified an emerging crisis in public opinion regarding contraception. Birth control was increasingly practised and accepted by the public (for personal and preventive reasons) but this position was at odds with the medical professional and government bodies that would provide it. The already complex working relationship between these groups was at risk of worsening because each group assumed the others were responsible for population issues. The official reticence towards contraception stifled individual need because the public, unaware that 'doctors [we]re public servants and the Ministry of Health a public machine', were too ashamed and timid to 'demand [the] help' they were due.[85] The medical profession assumed the state to be responsible for effecting and teaching preventive health strategies like contraception, and contended individuals were ultimately personally responsible for implementing those lessons. The doctor's job was treatment, not prevention.[86] Similarly, the state asserted its recent legal provisions provided the means for individuals to access medical and educational services through local health authority clinics, where medical health officers were empowered to effect large-scale preventive sanitary and welfare programs. If these services were not having the desired effect, it was the fault of the medical profession or populace.[87] In each scenario, the public, particularly women, bore the ultimate blame for failure, as they tended to be unaware of or failed to demand preventive sanitary and health education services and programs.[88] Even those who learned enough to avail themselves of technical and medical advice were chastised that their own 'defective mothering and housekeeping', not poverty, was the prime cause of illnesses.[89] Thus, the public rightly perceived that though medical and government authorities were somewhat responsible for public health, both groups were largely unsympathetic to public requests for contraceptive education and supply even as a preventive health tool, and 'prefer[red] to trust gossip'.[90]

This communal evasion of responsibility for contraceptive health led to the realisation of both the fears and hopes of the medical committee of the 1927 National Birth-rate Commission. It facilitated a booming contraceptive industry dominated by charlatans and peddlers 'whose only purpose [was] the sale of some particular apparatus' or compound to profit off the ignorance 'of those married couples who desire information as to contraceptives'. But, more importantly, forced the formation of contraceptive instruction clinics 'under the guidance and control of experienced and judicious medical practitioners'.[91] 'Far-seeing and generous individuals' operating on a voluntary basis banded together, amassed resources, and 't[oo]k action on their own initiative' to open and operate such clinics in response to official indifference to a legitimate and urgent public need.[92]

The NKWWC and all early contraceptive clinics were placed in a uniquely powerful and simultaneously weak position. Their scope for self-determination and direction was absolute; however, their reach was limited by the availability of funds, resources and volunteers dedicated to providing the service.[93] As most of these sites were established to meet desperate need in destitute areas, services were provided at minimal cost, and often free of charge. However, as these sites and services became better known after 1935 through the NBCA/FPA, large clinics like the NKWWC introduced a sliding scale of remuneration for private contraception patients, based on their reported weekly income. This was to check the practice of middle-class women attending contraceptive clinics to capitalise on the affordable medical expertise and advice as well as discounted contraceptives available therein.[94] Charging women who could afford to pay more for these services increased clinic revenue that could be funnelled into research and education programs that would ultimately benefit all future patients.[95] But, most extant voluntary health services that aimed to assist women, children and infants, the elderly, unemployed, chronically and mentally ill, and destitute, were bound by their capacity to attract subscriptions and donations, even after the Local Government Act (1871), which enabled local health authorities to service these communities under the Local Medical Officer of Health's direction.[96]

As with the two earlier clinics, Marie Stopes' Mothers' Clinic and the Malthusian League and later SPBCC-led Walworth Women's Welfare Centre (WWWC), the NKWWC heavily relied on voluntary funds and behaviours to operate. It was manned entirely by unpaid predominantly female staff from its directors, to the physicians and nurses, and ultimately the lay staff that managed the function and administration of the site. The latter also undertook initial patient consultations to facilitate the function and management of SPBCC clinics; this was a practice that Stopes reviled as 'very unfortunate' as it somewhat diminished the medical veneer that was then being applied to contraception via clinics.[97] Despite this concern, the NKWWC's use of committed lay female staff, especially 'sympathetic, kind, and patient' women as the first point of in-clinic contact proved an effective method of swiftly eliciting comfort and candour from patients. It was effectively employed to foster a sense of fellowship in their collective experiences as wives, mothers, and current/prospective contraceptive users.[98]

Voluntary clinic staff facilitated the centre's medical and contraceptive sessions. These women oversaw the constant cleanliness and comfort of the waiting room, ensured minor children were supervised during their mothers' medical consultations, and managed the adequacy of the clinic's contraceptive, medical, promotional and administrative supplies. As the NKWWC grew, a layperson was appointed as Clinic Superintendent to manage these staff: 'she [was] in charge at every session' and oversaw and directed all non-medical aspects of clinic function.[99] In the mid-1930s the clinic superintendent became responsible for judging if patients fell under the maximum £5 income threshold the executive committee had set for the receipt of reduced fees, and where individual patients fit on the sliding payment scale, and charging them accordingly.[100]

In 1931 Evelyn Fuller, superintendent of the Walworth clinic, published *On the Management of a Birth Control Centre*. This was a guide for contraceptive clinic design and operation and is a testament to the refinement of clinic management even before the NBCA had formalised its methods and practices. The publication reveals the urgent expansionist goals of early association affiliates, as no guide would be needed if there was no agenda to found clinics in remote localities, distant from the direction and advice of experienced staff.

Using the Walworth and North Kensington centres as a model, Fuller's guide defined a verified protocol for structuring and running clinics and sessions and provided a comprehensive list of essential equipment. Standardised style, height, design and comfort requirements were already set for SPBCC clinics, having originally been established at the WWWC and confirmed at the NKWWC. These designs had proved satisfactory at both sites and ensuing guidelines for regional clinic construction and function were based thereon. The NBCA mandated that affiliated clinics must contain a sizeable waiting room to accommodate patients and their children, in which all administrative functions were also managed. The area required sufficient room for storing patient files and data, contraceptives and other necessary tools of the lay staff as well as space to construct dressing facilities, for patients to prepare for medical examination.

Fuller's manual also specified the practical consulting room arrangements. She defined the minimum number of cubicles and the required floor space and necessary features for privacy and medical examination. Each cubicle required a gynaecological couch with specified dimensions topped with a thick mattress and pillow both dressed in 'white American cloth' and removable 'rubber sheeting' for sanitation. Finally, each cubicle required an 'electric light suspended from the ceiling' with a pulley attachment to accommodate the doctor's inspection.[101] These specifications had by 1931 been revised and proven during over eight years' clinical application, and this publication marked the point when the NBCA and its affiliates were prepared to confidently set and export standards for the design, management and function of affiliated clinics, and articulate the roles and responsibilities of its day-to-day lay and medical administration. After 1933 anyone approaching the association to establish a clinic was informed that any sites associated with and approved by the NBCA 'are expected to conform to certain [established] standards'.[102] This was always intended to be an expansionist project with the association at the national helm.

In 1934 Marie Stopes' published a pamphlet *Equipping a Birth Control Clinic*, it supported Fuller's clinic standards whilst simultaneously belittling them as 'vague and excessive ideas [regarding] the difficulty and amount of equipment necessary to establish a birth control clinic'.[103] Stopes, Britain's preeminent birth control advocate, educator and sexologist, had by now spent almost two decades studying sex, reproduction and marriage and published extensively and authoritatively on these matters. From her Mother's Clinic in North London she became the first Briton to disseminate contraceptive products and teaching in a medical environment and launched the Society for Constructive Birth Control and Racial Progress to facilitate a broad extension of her practice. In 1923, two years after opening her site, Stopes published *Contraception* (1923) a birth control textbook, moved her clinic to a more central location, opened some auxiliary branches and instituted a travelling caravan clinic. But as has been demonstrated by Claire Debenham, Caroline Rusterholz and Claire Jones, her fame quickly turned to eccentricity and notoriety and her star and influence waned.[104] She became 'arrogant, irrational, unco-operative and megalomaniacal' and distanced her clinics and self from the British contraceptive movement, which Peter Nuushul contends 'threatened to dilute her efforts to promote [contraceptive] technology'.[105] Ultimately this meant that she was excluded from the movement's post-1930 achievements, owing to her unwillingness to concede authority or cooperate effectively with others who shared her contraceptive goals. Her competitive rather than cooperative spirit explains why Stopes included snide remarks in her pamphlet to demonstrate the superiority of her clinical approach at the expense of what she called the 'very unfortunate ... Walworth technique'. Despite her claims, Stopes' recommendations reinforced Fuller's, with minor exceptions such as favouring nurses over doctors and disallowing interactions between lay workers and patients.[106] So, despite Stopes' bluster and seeming opposition, the NBCA method became supreme and informed the foundation of clinics around the nation.

Following the standardisation of general foundational and functional aspects of clinic programs and spaces, the NKWWC lay and medical committees concentrated on establishing a clinical atmosphere and medical management style. Despite lay staff having control over clinic administration, the medical committee retained an interest in streamlining and improving work conducted by even non-medical staff wherever possible. The committee determined that formalising and standardising clinic practices was integral to the NBCA achieving its ultimate goal to medicalise contraceptive practice and prescription and so, great attention was paid to presenting a formal medical appearance to the public. Behind the scenes, it was stressed that this was a key goal towards achieving legitimacy for the NBCA.

NKWWC and indeed all NBCA clinic sessions attempted to replicate typical doctors' appointments. The importance of this would likely have been lost on many early patients whose acquaintance with regular clinic practice would have been limited or non-existent owing to the limits of the National Insurance (1911) scheme, which only extended medical attention to women in relation to their maternal role. Thus, this was a self-legitimising exercise intended to project and ingrain a medical attitude, rather than to convince the few members of the public that came in the early years of its medical credentials. Upon arrival the patient would be greeted and some specifics relating to her social and financial circumstances ascertained. Following an ideally short wait in the designated waiting area, a doctor would attend to the patient. In private conference the physician would complete the patient's case card by conducting a medical survey querying her general health, past marriages and pregnancies, miscarriages and confinements, the overall health and survival rate of her children, methods and effectiveness of contraception employed currently and historically, and the acceptability of each method to both partners. This would be followed by a gynaecological questionnaire, probing her menstrual health and regularity, specifically monthly duration, blood loss and dysmenorrhoea, as well as dis/comfort during sexual intercourse, and urinary and bowel movements. Then the doctor would conduct a pelvic examination both by hand and with a speculum.[107] Often, this assessment discovered 'latent internal trouble which might never have been detected until harm was done'. Initially such patients were referred to a gynaecologist or hospital for treatment before further contraceptive consultation or fitting, but perceiving a potential market for the creation of expertise and patient loyalty, the NKWWC quickly introduced a treatment schedule for minor gynaecological issues to streamline the process.[108]

For patients without gynaecological issues, the doctor would determine and prescribe an appropriate cap type and size by measuring the depth of the patient's vagina, the tone of the perineum and condition of the pelvic organs, as well as a complementary spermicide.[109] The NKWWC took pains to stress there was 'no routine contraceptive technique at the clinic' and that its doctors prescribed a personalised method based on each patient's individual conditions and needs.[110] This was a not-so-hidden jab at Stopes who was throughout her career wholly dedicated to the same 'pro-race' cap and greasy suppository method.[111] After the medical assessment and prescription, the clinic nurse would offer 'detailed instruction in the technique of insertion and removal of the cap'. Once she was satisfied the patient could independently undertake both, a practice cap was issued, and the patient was instructed to return the following week wearing the cap for the doctor to examine her proficiency. If the patient was found competent and reported no irritation, a cap and instruction card was issued.[112]

The NKWWC was committed to prescribing women contraceptive technologies that allowed them to control their fertility and sexual health. The clinic's research and therapeutic focus was on 'an occlusive principle and a spermicidal principle' as they required 'no cooperation from the man'. The medical staff instituted what Rose Holz describes as a 'new kind of ritual' wherein the doctor, vaginal examination, contraceptive cap and follow-up visits were each prioritised in contraceptive prescription to ensure each patient had the capacity to effectively control her fertility.[113] This protocol was retained into the 1960s.[114]

During its first two decades the NKWWC attempted to service between twenty-five and thirty patients at each two-hour clinic session. However, in 1944 this was revised down to a maximum of twenty-five women.[115] This was necessitated by the formalisation of gynaecological examinations during *all* birth control sessions. This amplified demands on the doctor's time and the duration of each already busy appointment. Another important medical program introduced was 'special gynaecology sessions' for women whose doctors had 'discover[ed] an unhealthy cervix'.[116] This new service realised sessions as being intrinsically diagnostic and therapeutic at every level of practice.

The British voluntary birth control movement through its clinics provided the initial source of medical interaction for a large, typically low, class of women who were tacitly excluded from health services under every branch of state-organised health provision. The National Insurance system exclusively serviced taxpaying men; and the maternity benefits scheme introduced in 1913 offered worker's wives only limited health care as child bearers.[117] This produced a significant population of women who failed to qualify for state-funded health care and those in desperate need had to rely on the remnants of the Poor Law medical system for infirmary-based treatment.[118] Thus free (primarily reproductive) health centres like the NKWWC provided a safeguard against unnecessary debility and death to women who demonstrated a 'general disinclination to fuss about themselves'.[119]

Introducing Specialist Medical Services and Standards

The North Kensington establishment and environment was undoubtedly medical by the 1930s. The presence of the medical committee to mediate and direct the clinic translated into a sustained drive to establish scientific evidence that the NKWWC's products and methods were the safest, most effective and acceptable. Practically, standardisation meant doctors could quickly and efficiently evaluate patient's physical and mental ability in order to tailor contraceptive and medical advice. The primary figure leading the local medicalisation process was Chief Medical Officer Helena Wright. She pioneered the introduction and later compulsory application of gynaecological inspection for contraceptive provision. And in collaboration with other affiliated medical women, a huge volume and scope of specialised medical services broadly relating to sex were introduced through the site and later its branch clinics in Hounslow and Edgware. In the 1930s, Kathleen Harding introduced a sub-fertility and infertility focus and from 1945 a conception clinic, and Margery Spring Rice and Joan Malleson founded education and marriage difficulty sessions between the 1930s to 1950s.

Wright favoured a laissez faire approach to clinic medical treatment, where rather than being the exclusive patient of the clinic doctor that had first attended her, 'patients should have every light thrown on fitting and see as many doctors as possible'.[120] This was a modern approach to the traditional doctor–patient relationship, and reflected the patient-as-a-person movement that was currently in vogue.[121] She also demonstrated deep understanding of the importance of patients in achieving the association's contraceptive aims. In an era when, despite the decreasing birth rate, contraceptive knowledge, practice and even commercial publicity remained limited and professionalised, clinic success relied on repeat custom and good word-of-mouth; thus like general physicians in the nineteenth century, NKWWC doctors had to satisfy patient needs and desires.[122] The association knew that former patients made 'the best propagandists' and tried to court positive publicity by providing diverse programs to meet clients' various health needs.[123] This concern with patient satisfaction proved premonitory, as within a decade patient retention became an issue at association clinics, and the NKWWC pioneered a series of studies designed to increase patient trust and encourage semi-annual return visits.[124]

The increased pace of the medicalisation of sex, fertility and contraception from the early twentieth century greatly aided the NBCA's efforts, however, legitimation via state and medically sanctioned reproductive aftercare was vital to effect the association's broader women's health agenda. So, the Ministry of Health's sanctioning of contraceptive therapy for patients with gynaecological impairment for whom pregnancy presented an unacceptable risk, galvanised Wright to advise the NKWWC to introduce a gynaecological clinic. This astute addition to the NKWWC and NBCA repertoire effectively legitimised the clinical nature of women's welfare clinics, formally allying them with state- and medical-led preventive reproductive health programs. This new service both legitimised and standardised preventive gynaecological therapy under NBCA clinic and physician control.

From 1931 the NKWWC instituted gynaecology clinics with an initial goal of assessing the potential need and scope for this service.[125] It aimed to 'facilitate the early detection and prompt treatment of minor gynaecological ailments, which may develop, if untreated, into serious illness'.[126] In *Metropolitan Maternity*, Marks contends that this service was instituted as a means to advise and treat women excluded from state provided medical care.[127] But it had a further legitimising purpose that Caroline Rusterholz has identified, it formally aligned the medical provision at North Kensington within the acceptable specialties and noted expertise of female physicians: women's and community health.[128] According to the NKWWC records trial, demand was so high that the 1933 patient return rate was 80 per cent.[129] These sessions which likely constituted the patient's first post-natal medical care, allowed women to be thoroughly scrutinised by a doctor with specific gynaecological training adept to 'recognise the symptoms ... [the patient] probably does not'.[130] The addition of this service facilitated the gyanecologist to provide medical and pharmacological prescription or refer serious cases for specialised attention.[131] To bolster and inform Rice's mid-1930s Women's Health Enquiry which found almost half of the 1250 participating working women suffered serious gynaecological ailments,[132] North Kensington introduced weekly gynaecological examination sessions.[133] Attendance this decade was consistently high, compelling the site to schedule more weekly sessions. In 1933, when the Minister of Health George Newman conceded that a specialist gynaecology service was lacking in the National Health Service, the association could boast that it already provided an

> institution under adequate and skilled medical supervision [that] could receive minor gynaecological cases, deal with some forms of post-natal after-care, counsel... mothers of subnormal physique or mentality, [and] give advice on the practice of contraceptive methods when medically needed.[134]

The NKWWC and NBCA aimed to comprehensively provide these services. The pursuit was encouraged by the Departmental Committee on

Maternal Mortality and Morbidity's findings that advised mass gynaecological observation and oversight, as well as statistics derived from the first 1000 NKWWC patients which demonstrated only 158 were gynaecologically healthy.[135] Tackling these disproportionate debility rates provided the association an opportunity to directly incorporate government and medically sanctioned therapy as a standard aspect of clinic service. The NBCA manipulated its clinical observations and researches to claim that gynaecological health provided a tangible baseline to predict maternal, infant and family conditions; and used this to inspire attendance and dependence on its clinics to address marital and maternal health issues, and trust in treatments it offered.

The NKWWC, it was claimed, would 'strike at the root of maternal ill health' by teaching women 'to learn the care [for] their own health before ... childbearing'.[136] The clinic guaranteed each woman a 'private consultation... with a woman doctor', who would use a vaginal speculum to conduct a thorough pelvic examination and identify any abnormality.[137] The British birth control industry was then mostly populated by women and so advertising this served to both notify potential clients of that fact, and articulate a sense of knowing about the female body and its reproductive and sexual function and health.[138] Providing this service also offered the association further means to evaluate and guarantee contraceptive efficacy, as certain gynaecological ailments impacted contraceptive ability and comfort. Gynaecology clinics thus provided association physicians the opportunity to better understand and assess women's reproductive health to effectively identify and treat issues and ensure apt contraceptive prescriptions.

The success and popularity of the NKWWC's gynaecology sessions led to the inclusion of remedial gymnastics as a treatment option in 1950. This program initially trialled under Margaret Kirschner at the Uxbridge auxiliary clinic,[139] 'spread the idea that the body can be put or, better still, kept in order through one's own efforts'. The treatment was essentially a form of gynaecological physiotherapy to treat patients suffering from 'slack perineum, prolapsed vaginal walls, extreme mobility of [the] uterus; slack abdominal walls (especially after Caesarian) [and] Lordosis'.[140] This offered a form of treatment through which 'suitable patients [could] avoid the necessity of wearing [internal] surgical appliances' through self-focus and bodily regulation. Kirchner reported the method had proved 'fairly successful' for advanced uterine prolapse patients and intended to offer group sessions for maintaining good 'habits of muscular co-ordination'. Remedial gymnastics was rotated through auxiliary clinics and publicised through a series of lectures and recruitment drives amongst local health authorities, hospitals and voluntary societies in the early 1950s.[141] The proposed introduction of collective physiotherapy sessions indicates that, as Edith Macrae has shown, group fitness classes were not out of the norm for adults by this time; however remedial gymnastics classes that were medically focused and prescribed did not become a preventive cultural phenomenon as was hoped, but remained a niche service associated with pre and antenatal health issues.[142] Unfortunately by 1953, the 'exercise clinic was [deemed] very slack [as] sometimes no patients attended and often only two or three'. Within a few years, the service was wound down as increasing numbers of health visitors and midwives became 'trained in pre and post-natal exercises'.[143]

To complement its gynaecological focus, the NKWWC also introduced a program of minor surgical intervention to tackle common complaints. In 1955 physician Mary Pollock introduced two surgical procedures to the gynaecological repertoire: 'cauterization of cervical erosion' and 'excision of hymenical bands, or incision of rigid hymens under local anesthetic'.[144] These operations promised to streamline clinic services to ensure expedient local treatment without needing to refer patients for specialised care as was protocol. It was further argued that carrying surgical instruments would 'be useful in dealing with any bleeding that might occur during examinations'.[145] Surgical interventions were a logical extension of association clinics services and this, like its entire gynaecological focus, was used to advance the 'constructive, qualitative nature of [clinic] work' that Leathard claims followed the 1939 rebranding of the association to focus on family planning rather than birth control.[146] This more invasive service prompted another preventive health innovation, the introduction of a pathological examination schedule that collected and assessed patients' vaginal and cervical smears, biopsies and urine.[147] By the mid-1950s the principle of prevention and early diagnosis for cancer was well established, and owing to long-standing rumours and concerns that contraception was a prime cause, such examinations became a keen local and then national focus for the association.

Late 1930s North Kensington noted an increase in returning patients, both to the clinic and its now well-established bi-weekly gynaecology sessions. This indicated mounting public knowledge and appreciation of a definite relationship between gynaecological and general health as well as a perception that marital strength and sexual harmony were fundamental to wellbeing. This insight prompted the clinic to implement both educational and psychological marital advice sessions, for women initially, but later expanding to service men too. This demonstrates that the interested and informed public increasingly relied upon NBCA clinics and staff for medical, contraceptive and educational guidance to achieve personal marital, sexual and family ambitions. Increased association patronage and trust was undoubtedly bolstered by the implementation and success of medically and scientifically verified methods and technologies in therapeutic practices, and as a result produced patients willing to receive instruction and advice in other aspects of marriage.

In 1936, Rice initiated sex education talks for married patients to address 'any matter which perplexed' them.[148] By April 1937, these talks, addressed exclusively to 'Women and Girls', proved so popular that the NKWWC executive committee requested they become a weekly clinic feature.[149] Two years later, Malleson, a sex and marriage counsellor, overtook the sessions and also instituted a 'difficulties of marital adjustment' clinic to help women to adapt to the social and sexual aspects of marriage. Concurrently, the association suggested that 'men's difficulties session' might also be merited[150] and employed psychologist Douglas Bryan to design and administer a program specifically for male clientele.[151] Between 1939 and 1951 he ran the NKWWC male sexual difficulty program and pioneered a program of therapy sessions for men, women and couples.[152]

After World War Two, interest in sexual and marital difficulty and harmony became a keener focus for Britons and as a result demand for guidance from the FPA and NKWWC expanded significantly. In response, in 1948 North Kensington held an association-wide conference to encourage other associations to assume psychological and medical sex and contraceptive therapies.[153] As Rusterholz has shown, association-affiliated Wright, Malleson and Mary Macaulay and other prominent medical women widely presented their clinical experience addressing sexual difficulties, psychosexual development and social marital adjustment problems in private and in clinic.[154] Marriage guidance medico-psychologist Noel Harris and psychotherapist Stella Churchill also articulated successful psychological therapies they had each employed.[155] The interest these services generated inspired the NKWWC to expand its marital difficulty clinics and therapeutic staff over the next few years.[156] In the early

1950s, a split occurred between the association and NKWWC regarding the therapeutic value of education and counselling in association clinics, and the former attempted to silo these services to non-medical societies like the Family Relations Group and Marriage Guidance Council (see Chapter 3). This marked the first major disagreement between the two regarding medical and scientific services provided at the site. But Rice and her medical advisors remained adamant counselling and education was medical and in 1951 planned to extend marriage guidance programs to include the unmarried.[157] This expansionist goal was unfulfilled, as over the next two decades scientific and medical contraceptive and fertility studies took priority.[158]

In the mid-1930s sub- and infertility became an association interest owing to growing numbers of patients approaching clinics for conception assistance. At the time questions of sub- and hyper-fertility were beginning to occupy social, political and medical consciousness and a national population crisis was believed imminent. Anxieties about fertility and demographic change as the second global conflict concluded, prompted the institution of a Royal Commission on Population that explicitly investigated the scope and types of contraception being employed in contemporary Britain and considered whether it was having a negative effect on national fertility.[159] Just prior in 1943, the FPA convened a sub-fertility committee to determine if 'involuntary sterility' was increasingly presenting at clinics, and if there was a need to introduce 'Motherhood Clinics'.[160] The NKWWC responded 'yes' to both and its medical committee initiated half hour sub-fertility sessions by appointment and charged patients who met sub-fertile criteria a small fee for fertility testing.[161] This relaxed appointment-based system continued for almost two years until the publicity surrounding the Royal Commission's prompted a formalisation of this work.

In February 1945 the NKWWC appointed Kathleen Harding its specialist gynaecologist in charge of sub-fertility.[162] In that role she worked closely with fertility experts to refer patients needing specialised care and testing which neither the centre, nor association, could then accommodate. Sub-fertility testing for men was obviously far easier than for women and could be undertaken quickly and easily by the association's scientific researchers. Despite the straightforwardness of male testing, requests for the spermological service almost always followed Harding's determination of the wife's fertility through the results of tubal insufflation tests and 'lipiodal investigation' showing no tubal obstruction.[163]

This clearly indicated that fertility was, at least culturally, the woman's domain and that prospective mothers would rather financially and physically exhaust all female options before questioning their husband's virility. Still, demand was high enough for the association to establish a dedicated Seminological Laboratory in 1943 under H. A. Davidson to test sperm quantity and motility to identify if the husband was indeed the cause of a couple's involuntary sterility.[164] Davidson accepted all referrals, as he had recently collected some concerning statistics regarding male sub-fertility. During the laboratory's formative years, he had treated '710 patients, 540 male and 170 females (for post coital tests)'; 181 had been referred by FPA clinics and Hospitals with most coming directly from North Kensington. Of the males tested, 10 per cent were sterile, 47 per cent definitely sub-fertile, 14 per cent borderline cases and only 29 per cent were fertile.[165] Rusterholz discusses the pathologisation of infertility as female failure, but as Davidson's fertility results show—this was an unfair medical bias that was not borne out by his findings.[166] The 1950 figures produced by the NKWWC reported that approximately twice as many women (99) as men (53) presented for sterility tests, of them 'approximately 50 per cent' of the husbands were deemed sub-fertile and they were the sole cause of 15 out of 24 cases where 'no pregnancy was expected'. Approximately sixty per cent of wives after being tested were proved to have no physiological cause for sterility, but about forty per cent presented with either blocked fallopian tubes or tubal spasm.[167] As the century progressed and greater numbers of couples and individuals presented for these services it was increasingly accepted that both men and women could be and were the cause of infertility and that the field was a worthy association focus. So mid-century Wright pushed and Harding agreed to have all 'inexplicable failures' to conceive that presented at the NKWWC promptly fertility tested.[168] There was also a concerted effort for the clinic to become the FPA's prime sub-fertility provider as none of the smaller sites had room to accommodate the service.[169] This eventuated, as did an expansion, and by 1957 twelve FPA clinics hosted dedicated sub-fertility sessions, and 170 offered advice and referrals for specialised service.[170] The FPA endeavoured, relatively unsuccessfully, to attract greater medical attention to involuntary sterility through its clinical research and practice and after 1953 incorporated artificial insemination into its services.[171]

Conclusion

By the NKWWC's twenty-ninth year it had treated the contraceptive, gynaecological, psychological, sexual, marital and fertility concerns of over 46,000 patients; it also provided medical and lay contraceptive training sessions and referrals for specialised offsite therapy. Its executive committee directed the site; and 'all the senior doctors on staff' peopled the medical committee. In 1953, the NKWWC employed twenty-four sessional female doctors and sixteen nurses, in addition to a 'gynecologist, an endocrinologist, a medical statistician and specialists in sexual disorders' as consulting volunteers. Finally, five full-time and sixteen part-time lay staff managed its administration and record keeping.[172] The Centre opened all day from Monday to Friday, and ran sixteen contraceptive, gynaecology, marriage problems and involuntary sterility sessions weekly at North Kensington and from its auxiliary locations in Acton, Battersea, Edgware, Hayes, Hampton Wick, Hounslow and Uxbridge.[173] This clinic as the association's standard model, most certainly represented an ideal structural, service and practical peak.

Notes

1. Rose Holz, *The Birth Control Clinic in a Marketplace World* (Woodbridge: Boydell & Brewer, University of Rochester Press, 2012); Cathy Moran Hajo, *Birth Control on Main Street: Organizing Clinics in the United States, 1916–1939* (Urbana: University of Illinois Press, 2010); Lara Marks, *Metropolitan Maternity: Maternal and Infant Welfare Services in Twentieth Century London* (The Netherlands: Rodophi B. V. Amsterdam, 1996); Lara Marks, *Sexual Chemistry: A History of the Contraceptive Pill* (London: Yale University Press, 2001); Audrey Leathard, *Fight for Family Planning: The Development of Family Planning Services in Britain, 1921–74* (London: The Macmillan Press, 1980); Hera Cook, *The Long Sexual Revolution: English Women, Sex and Contraception, 1800–1975* (New York: Oxford University Press, 2004); Beryl Suitters, *Be Brave and Angry: Chronicles of the International Planned Parenthood Federation* (London: International Planned Parenthood Federation, 1973); Clare Debenham, 'Grassroots Feminism: A Study of the Campaign of the Society for the Provision of Birth Control Clinics, 1924–1938' (PhD, University of Manchester, 2010); Caroline Walker, 'Making Birth Control Respectable: The Society for Constructive Birth Control and Racial Progress, and the

American Birth Control League, in Comparative Perspective, 1921–1938' (PhD, University of Bristol, 2007); Richard Soloway, *Demography and Degeneration: Eugenics and the Declining Birthrate in Twentieth Century Britain* (Chapel Hill: University of North Carolina Press, 1995); Adele E. Clarke, *Disciplining Reproduction: Modernity, American Life Sciences, and the 'Problems of Sex'* (Los Angeles: University of California Press, 1998); Ann Farmer, *By Their Fruits: Eugenics, Population Control, and the Abortion Campaign* (Washington: Catholic University of America Press, 2008); Richard Carr and Bradley W. Hart, *The Global 1920s: Politics, Economics and Society* (London: Routledge, 2016); Julia Stonehouse, *Idols to Incubators: Reproductive Theory Through the Ages* (London: Scarlet Press, 1994).

2. Barbara Evans, *Freedom to Choose: The Life and Work of Dr Helena Wright, Pioneer of Contraception* (London: The Bodley Head, 1984), 132.
3. Sylvia Dunkley, 'Rice, Margaret Lois Spring (1887–1970)', *Oxford Dictionary of National Biography*, accessed 1 July 2022, https://www.oxforddnb.com/view/10.1093/ref:odnb/9780198614128.001.0001/odnb-9780198614128-e-74760?rskey=DSVb3l&result=1.
4. Dunkley, 'Rice, Margaret Lois Spring'.
5. *The Labour Woman* 1, 8 (1913): 139. Quoted in Marks, *Metropolitan Maternity*, 275.
6. Christine Collette, *For Labour and For Women: The Women's Labour League, 1906–1918* (Manchester: Manchester University Press, 1989), 113.
7. Dorothy Porter, *Health, Civilization and the State: A History of Public Health from Ancient to Modern Times* (London: Routledge, 1999), 184–5.
8. James Fenton and the Royal Borough of Kensington, *The Annual Report of the Medical Officer of Health, 1924* (London: Vail and Co., 1925), 34.
9. Rough Notes 7 and 8 by Margery Spring Rice, [c.1928]. Wellcome Library, Archives of the Private Papers of Margery Spring Rice (WL/SA/SR), WL/SA/SR21.
10. Clare Debenham, *Birth Control and the Rights of Women: Post Suffrage Feminism in the Twentieth Century* (London: I.B. Taurus, 2013), 108.
11. Deborah A. Cohen, 'Private Lives in Public Spaces: Marie Stopes, the Mothers' Clinics and the Practice of Contraception', *History Workshop Journal* 35, 1 (1993): 97.
12. Marks, *Metropolitan Maternity*, 149.
13. Caroline Rusterholz, *Women's Medicine: Sex, Family Planning and British Female Doctors in Transnational Perspective, 1920–70* (Manchester: Manchester University Press, 2020), 4.
14. D. Porter, *Health, Civilization and the State*, 182–3.

15. D. Porter, *Health, Civilization and the State*, 136.
16. Sanitary Institute of Great Britain, *Transactions of the Sanitary Institute of Great Britain, Volume VII* (London: Offices of the Sanitary Institute and Edward Stanford, 1887), 68.
17. James Fenton and the Royal Borough of Kensington, *The Annual Report on the Health of the Borough for the Year 1925* (London: Vail and Co., 1926), 86.
18. Marks, *Metropolitan Maternity*, 149.
19. Johann Ferch, *Birth Control* (London: Rider and Co., 1932), 11.
20. Ministry of Health, *Memorandum 153/M.C.W.* (London: Ministry of Health, July 1930); Ministry of Health, *Circular 1208* (London: Ministry of Health, 1931).
21. Ministry of Health, *Memorandum 153/M.C.W*.
22. Marks, *Metropolitan Maternity*, 149; Richard Soloway, *Birth Control and the Population Question in England, 1877–1930* (Chapel Hill: University of North Carolina Press, 1982), 315.
23. Helena Wright, *Birth Control: Advice on Family Spacing and Healthy Sex Life* (London: Cassell's Health Handbooks, 1935), 76.
24. Soloway, *The Population Question*, 315.
25. 'Medical News', *British Medical Journal* 2, 3688 (1931): 516.
26. Ministry of Health, *Circular 1208.*
27. Ministry of Health, *Interim Report of Departmental Committee on Maternal Mortality and Morbidity* (London: His Majesty's Stationery Office, 1930), 44.
28. Ministry of Health, *Final Report Departmental Committee on Maternal Mortality and Morbidity* (London: His Majesty's Stationery Office, Wyman and Sons, 1932), 130–1.
29. Ministry of Health, *Circular 1408* (London: Ministry of Health, 1934).
30. Dunkley, 'Rice, Margaret Lois Spring'.
31. Marks, *Metropolitan Maternity*, 148–9, 277–84; Audrey Leathard, *Fight for Family Planning*, 57.
32. Norman E. Himes and Vera C. Himes, 'Birth Control for the British Working Classes: a Study of the First Thousand Cases to Visit an English Birth Control Clinic', *Hospital Social Service* 19 (1929): 582.
33. Gladys M. Cox, *Clinical Contraception*, Second Edition (London: William Heinemann [Medical Books] Limited, 1937), 182.
34. Marks, *Metropolitan Maternity*, 277–80.
35. NKWWC Seventeenth Doctors Committee Report, 2 August 1934. Wellcome Library, Archives of the Family Planning Association (WL/SA/FPA), WL/SA/FPA/NK224.
36. *NKWWC Annual Report, 1928–9*, 1929. WL/SA/SR7; Evans, *Freedom to Choose*, 133–4.

37. 'Birth Control and Contraception', *Journal of the American Medical Association* 103, 15 (October 1934): 1169.
38. Clarke, *Disciplining Reproduction*, 192.
39. Humphry Rolleston, 'Birth Control Investigation Committee', *British Medical Journal* 2 (1927): 805–6.
40. NKWWC Seventeenth Medical Committee Report, 2 August 1934. WL/SA/FPA/NK224.
41. Haven Emerson, Wilbur Ward and Frederick C. Holden, 'Birth Control and Contraception', *Journal of the American Medical Association* 103, 15 (1934): 1169.
42. NKWWC Medical Committee Report, 15 November 1934. WL/SA/FPA/NK225.
43. NKWWC Medical Committee Report, 15 November 1934. WL/SA/FPA/NK225.
44. Emerson, Ward and Holden, 'Birth Control and Contraception', 1169.
45. NKWWC Medical Committee Report, 24 October 1934. WL/SA/FPA/NK225.
46. Emerson, Ward and Holden, 'Birth Control and Contraception', 1169.
47. Claire L. Jones, *The Business of Birth Control* (Manchester: Manchester University Press, 2020), 110–1.
48. NKWWC Medical Committee Report, 15 November 1934. WL/SA/FPA/NK225.
49. Cecil Voge, 'Discussion on Chemistry and Physics of Contraceptives', in *Biological and Medical Aspects of Contraception*, ed. Margaret Sanger (Washington: The National Committee on Federal Legislation for Birth Control, 1934), 119–30.
50. The first dated approved list is from October 1936. Thus, the idea was likely being actively considered but had yet to be actioned in November 1935. Approved List, October 1936. WL/SA/FPA/A7/1.
51. NKWWC Medical Committee Report, 24 October 1934. WL/SA/FPA/NK225.
52. NKWWC Medical Committee Report, 4 February 1935. WL/SA/FPA/NK225.
53. L. Jeger, 'The Politics of Family Planning', *Political Quarterly* 33, 1 (1962): 51.
54. National Birth Control Association, *NBCA Annual Report, 1935–6* (London: W. H. Taylor and Sons, 1936), 3.
55. NKWWC Medical Committee Report, 24 October 1934. WL/SA/FPA/NK225.
56. Graeme Gooday, 'Liars, Experts and Authorities', *History of Science* 46, 154 (2008): 431–456.
57. NKWWC Medical Committee Minutes, 4 February 1935. WL/SA/FPA/NK225.

58. NKWWC Medical Committee Report, 15 November 1934. WL/SA/FPA/NK225.
59. NKWWC Medical Committee Minutes, 24 October 1934. WL/SA/FPA/NK225.
60. NKWWC Medical Committee Minutes, 15 November 1934. WL/SA/FPA/NK225; NKWWC Medical Committee Minutes, 24 October 1934. WL/SA/FPA/NK225.
61. Claire Jones, '"Under the Covers?" Commerce, Contraceptives and Consumers in England and Wales, 1880–1960', *Social History of Medicine* 29, 4 (2016): 738–9.
62. Memorandum to Secretaries of Affiliated Centres, 14 January 1935. Wellcome Library, Archives of the Eugenics Society (WL/SA/EUG), WL/SA/EUG/D/12/32.
63. NKWWC Medical Committee Minutes, 15 November 1934. WL/SA/FPA/NK225.
64. NKWWC Medical Committee Minutes, 15 November 1934. WL/SA/FPA/NK225.
65. Robert B. Fireworker and Hershey H. Friedman, 'The Effects of Endorsement on Product Evaluation', *Decision Sciences* 8, 3 (1977): 576–83.
66. NKWWC Medical Committee Minutes, 15 November 1934. WL/SA/FPA/NK225.
67. Jones, *Business of Birth Control*, 44–5.
68. NKWWC Medical Committee Minutes, 4 February 1935. WL/SA/FPA/NK225.
69. NKWWC Medical Committee Minutes, 26 September 1935. WL/SA/FPA/NK225.
70. Merriley Borell, 'Biologists and the Promotion of Birth Control Research, 1918–1938', *Journal of the History of Biology* 20, 1 (1987): 51–87; Ilana Löwy, 'Defusing the Population Bomb in the 1950s: Foam Tablets in India', *Studies in History and Philosophy of Biological and Biomedical Sciences* 43 (2012): 583–93; Ilana Löwy, '"Sexual Chemistry" Before the Pill: Science, Industry and Chemical Contraceptives, 1920–1960', *British Journal for the History of Science* 44, 2 (2011): 245–74; Caroline Rusterholz, 'Testing the Gräfenberg Ring in Interwar Britain: Norman Haire, Helena Wright, and the Debate over Statistical Evidence, Side Effects, and Intra-uterine Contraception', *Journal of the History of Medicine and Allied Sciences* 72, 4 (2017): 448–67; Richard Soloway, 'The "Perfect Contraceptive": Eugenics and Birth Control Research in Britain and America in the Interwar Years', *Journal of Contemporary History* 30, 4 (1995): 637–64; Leathard, *Fight for Family Planning*, 106–7.

71. NKWWC Medical Committee Minutes, 30 November 1936. WL/SA/FPA/NK226.
72. NKWWC Medical Committee Minutes, 22 April 1937. WL/SA/FPA/NK226.
73. Wright, *Birth Control*, 88.
74. Margery Spring Rice, *Working-Class Wives: Their Health and Conditions* (London: Penguin Books Limited, 1939), 44.
75. James Merchant, ed., *Medical Views on Birth Control, with an Introduction by Sir Thomas Horder* (London: Martin Hopkinson and Co. Ltd., 1926), 115.
76. David Pyke, 'Obituary: Lord Horder of Ashford', *British Medical Journal* 2, 4939 (1955): 624.
77. Merchant, *Medical Views on Birth Control*, ix–xiv.
78. Merchant, *Medical Views on Birth Control*, 68.
79. *Offences Against the Persons Act 1861* (Great Britain); *Infant Life Preservation Act 1929* (Great Britain); House of Lords Debate, *Child Destruction Bill*, 12 July 1928, 71: 998–1000. Place of publication: Hansard, http://hansard.millbanksystems.com/lords/1928/jul/12/child-destruction-bill-hl.
80. Merchant, *Medical Views on Birth Control*, 66–9.
81. Leathard, *Fight for Family Planning*, 46–7.
82. Joan Malleson, *The Principles of Contraception: A Handbook for General Practitioners* (London: Victor Gollancz Ltd., 1935), 23.
83. Malleson, *The Principles of Contraception*, 19.
84. Wright, *Birth Control*, 76–81.
85. Wright, *Birth Control*, 80–8.
86. Marks, *Sexual Chemistry*, 116.
87. Steven Cherry, *Medical Services and the Hospitals in Britain, 1860–1939* (Cambridge: Cambridge University Press, 1996), 48–9.
88. Rice, *Working-Class Wives*, 46.
89. Cherry, *Medical Services and Hospitals in Britain*, 39.
90. Malleson, *The Principles of Contraception*, 19.
91. Charles Gibbs and Arthur Newsholme, *Medical Aspects of Contraception: Being the Report of the Medical Committee Appointed by the National Council of Public Morals in connection with the Investigations of the National Birth-rate Commission* (London: Martin Hopkinson and Co., 1927), 12.
92. Wright, *Birth Control*, 82–3.
93. Wright, *Birth Control*, 79.
94. NKWWC Medical Committee Minutes, 14 February 1936. WL/SA/FPA/NK226.
95. NKWWC Medical Committee Minutes, 28 November 1935. WL/SA/FPA/NK225.

96. Cherry, *Medical Services and Hospitals in Britain*, 68–9.
97. Marie Stopes, *Equipping a Birth Control Clinic*, Second Edition (London: Mothers' Clinic for Constructive Birth Control, 1952), 11.
98. Stopes, *Equipping a Clinic*, 10.
99. Evelyn Fuller, *On the Management of a Birth Control Centre*, Second Edition (London: Noel Douglas, 1931), 34.
100. NKWWC Medical Committee Minutes, 14 February 1936. WL/SA/FPA/NK226.
101. Fuller, *Management of a Birth Control Centre*, 25–33.
102. Letter NBCA Acting Secretary to F. Cullingham, 21 October 1933. WL/SA/FPA/A13/5.
103. Stopes, *Equipping a Clinic*, 4.
104. Claire Debenham, *Marie Stopes' Sexual Revolution and the Birth Control Movement* (UK: Palgrave Macmillan, 2018), 13; Rusterholz, *Women's Medicine*, 217; Jones, *Business of Birth Control*, 8.
105. Peter Neushul, 'Marie C. Stopes and the Popularization of Birth Control Technology', *Technology and Culture* 39, 2 (1998): 246.
106. Stopes, *Equipping a Clinic*, 10.
107. Helena Wright, *Contraceptive Technique: A Handbook for Medical Practitioners and Senior Students* (London: J&A Churchill Ltd., 1951), 42.
108. Fuller, *Management of a Birth Control Centre*, 22.
109. Holz, *The Birth Control Clinic in a Marketplace World*, 36; NKWWC Medical Supplement, [c.1930]. WL/SA/SR15/1–25.
110. Supplement for Medical Readers, [c.1928]. WL/SA/SR15/1–25.
111. Jones, *Business of Birth Control*, 109, 162.
112. NKWWC Medical Committee Minutes, 28 November 1935. WL/SA/FPA/NK225.
113. Supplement for Medical Readers [c.1928]. WL/SA/SR15/1–25.
114. NKWWC Medical Committee Minutes, 9 February 1953. WL/SA/FPA/NK230.
115. NKWWC Medical Committee Minutes, 18 October 1944. WL/SA/FPA/NK227.
116. NKWWC Medical Committee Minutes, 5 September 1960. WL/SA/FPA/NK230.
117. Marks, *Metropolitan Maternity*, 198.
118. Cherry, *Medical Services and Hospitals in Britain*, 59.
119. Rice, *Working-Class Wives*, 39.
120. NKWWC Doctors Committee Notes, 6 July 1934. WL/SA/FPA/NK224.
121. Roy Porter, *The Greatest Benefit to Mankind: A Medical History of Humanity from Antiquity to the Present* (London: Fontana Press, 1997), 682.

122. R. Porter, *The Greatest Benefit to Mankind*, 671.
123. *NKWWC Annual Report, 1925–6*, 1926. WL/SA/FPA/NK/206.
124. NKWWC Clinic Superintendents Report, 22 April 1937. WL/SA/FPA/NK226; NKWWC Medical Committee Minutes, 2 August 1935. WL/SA/FPA/NK225.
125. Gynaecological Clinics, [c.1933]. WL/SA/SR15/1–25.
126. Suggested Letterpress for the Proposed Propaganda Leaflet, 1953. WL/SA/EUG/D/154.
127. Marks, *Metropolitan Maternity*, 253–4.
128. Rusterholz, *Women's Medicine*, 138.
129. NKWWC Gynaecological Sub-Committee Minutes, 11 July 1933. WL/SA/FPA/NK211.
130. Rice, *Working-Class Wives*, 37.
131. NKWWC Gynaecological Sub-Committee Minutes, 31 October 1933. WL/SA/FPA/NK211.
132. Rice, *Working-Class Wives*, 37.
133. NKWWC Medical Committee Minutes, 23 May 1935. WL/SA/FPA/NK225.
134. Gynaecological Clinics, [c.1933]. WL/SA/SR15/1–25.
135. Gynaecological Clinics, [c.1933]. WL/SA/SR15/1–25.
136. *NKWWC Annual Report, 1933–4*, 1934. WL/SA/FPA/NK/206.
137. NKWWC Propaganda Leaflet, May 1953. WL/SA/EUG/D/154.
138. Rusterholz, *Women's Medicine*, 138.
139. NKWWC Medical Committee Minutes, 13 December 1950. WL/SA/FPA/NK227.
140. Letter M. Kirschner to H. Wright, 8 May 1950. WL/SA/FPA/NK227; Letter NKWWC to Ministry of Health Maternity and Child Welfare Department, [c.1951]. WL/SA/FPA/NK194.
141. Letter R. Peers to A. R. Buck, 25 February 1954. WL/SA/FPA/NK194; Letter M. Waters to R. Peers, 8 March 1952. WL/SA/FPA/NK194; Letter M. Kirschner to H. Wright, 8 May 1950. WL/SA/FPA/NK227; Letter NKWWC to Ministry of Health Maternity and Child Welfare Department, [c.1951]. WL/SA/FPA/NK194.
142. Edith Macrae, *Exercise in the Female Life-Cycle in Britain, 1930–1970* (London: Palgrave Macmillan, 2016), 136–9.
143. NKWWC Medical Committee Minutes, 18 February 1957. WL/SA/FPA/NK230.
144. NKWWC Medical Committee Minutes, 15 August 1955. WL/SA/FPA/NK230.
145. Gynaecology, Executive Decisions, 1951–9. WL/SA/FPA/NK23.
146. NKWWC Medical Committee Minutes, 15 August 1955. WL/SA/FPA/NK230; Leathard, *The Fight for Family Planning*, 68.

147. Development of Gynaecological Clinic, [c.1955]. WL/SA/FPA/NK23; Pathological Examinations for Gynaecological Clinic, [c.1955]. WL/SA/FPA/NK23.
148. Proposed NKWWC Development Notes, June 1936. WL/SA/FPA/NK226.
149. NKWWC Clinic Superintendent's Report, 22 April 1937. WL/SA/FPA/NK226.
150. D. E. Martin, 'Malleson, Joan Graeme (1899–1956)', in *Oxford Dictionary of National Biography* (Oxford: Oxford University Press, 2004), accessed 1 July 2022, http://www.oxforddnb.com/view/article/54690; Services offered to Practitioners by the NBCA Medical Sub-Committee, 31 January 1938. WL/SA/EUG/D/12/16/22.
151. NKWWC Medical Committee Minutes, 13 January 1939. WL/SA/FPA/NK226.
152. NKWWC Medical Committee Minutes, 14 November 1945. WL/SA/FPA/NK227; NKWWC Medical Committee Minutes, 22 May 1951. WL/SA/FPA/NK227.
153. NBCA Medical Sub-Committee Minutes, 25 October 1948. WL/SA/FPA/A5/88.
154. Rusterholz, *Women's Medicine*, 91.
155. NBCA Medical Sub-Committee Minutes, 24 January 1949. WL/SA/FPA/A5/88.
156. NKWWC Medical Committee Minutes, 8 September 1949. WL/SA/FPA/NK227; NKWWC Medical Committee Minutes, 10 August 1950. WL/SA/FPA/NK227.
157. Confidential Memorandum Margery Spring Rice to NKWWC Medical Committee, 1951. WL/SA/FPA/NK227.
158. NKWWC Medical Committee Minutes, 5 January 1953. WL/SA/FPA/NK230.
159. E. Lewis-Faning, *Papers of the Royal Commission on Population, Volume 1: Report on an Enquiry into Family Limitation and its Influence on Human Fertility During the Past Fifty Years* (London: His Majesty's Stationery Office, 1949).
160. NBCA Medical Sub-Committee Minutes, 30 March 1943. WL/SA/FPA/A5/88.
161. NKWWC Medical Committee Minutes, 26 May 1943. WL/SA/FPA/NK227; NKWWC Medical Committee Minutes, 30 September 1942. WL/SA/FPA/NK227.
162. NKWWC Medical Committee Minutes, 28 February 1945. WL/SA/FPA/NK227.
163. NKWWC Medical Committee Minutes, 26 May 1943. WL/SA/FPA/NK227.

164. NBCA Medical Sub-Committee Minutes, 10 October 1943. WL/SA/FPA/A5/88; Margaret Pyke, *Family Planning: An Assessment, Galton Lecture, 27 February 1963* (Mitcham: West Brothers Printers, 1963), 7.
165. NBCA Medical Sub-Committee Minutes, 17 June 1946. WL/SA/FPA/A5/88.
166. Rusterholz, *Women's Medicine*, 122–5.
167. NKWWC 1950 Subfertility Figures, 1951. WL/SA/FPA/NK/95.
168. NBCA Medical Sub-Committee Minutes, 8 April 1946. WL/SA/FPA/A5/88; NKWWC Medical Committee Minutes, 18 February 1947. WL/SA/FPA/NK227.
169. NKWWC Medical Committee Minutes, 5 January 1953. WL/SA/FPA/NK230.
170. Rusterholz, *Women's Medicine*, 123.
171. NKWWC Medical Committee Minutes, 3 April 1952. WL/SA/FPA/NK230.
172. NKWWC Propaganda Leaflet, May 1953. WL/SA/EUG/D/154.
173. FPA Clinics in the London Area, September 1952. WL/SA/EUG/C/282.

CHAPTER 3

Teaching and Networking the Wright Way

Abstract This chapter explores National Birth Control Association/ Family Planning Association Medical Director Helena Wright's joint efforts with the association and its official and unofficial networks to introduce comprehensive, scientific sex and contraceptive education for the public and medical professionals. Together they envisioned and worked to create a basic standard of sexual and contraceptive literacy for Britain. This chapter uses Wright's sexological texts, association and network syllabi, and formal and informal association, clinic, and Royal Commission on Population-related records to depict the formulation of standardised scientific sex and contraceptive education delivered to the public and medical community via lessons learned from the bedroom and contraceptive clinic.

As the broad agenda and programs of the North Kensington Women's Welfare Centre demonstrate, from the 1930s birth control clinicians, lay people and family-focused welfare networks worked independently and collectively to inform the public about sexual relations, fertility and health. From that time the National Birth Control Association (NBCA)/Family Planning Association (FPA) and its prolific medical and educational pioneering women, particularly Helena Wright, engaged with the public and medical professionals to provide increasingly standardised comprehensive scientifically sound, experience-based contraceptive

N. Szuhan, *The Family Planning Association and Contraceptive Science and Technology in Mid-Twentieth-Century Britain*, Medicine and Biomedical Sciences in Modern History, https://doi.org/10.1007/978-3-030-81300-0_3

instruction. Wright, through her medical and association-based experiences and networks, determined that the ultimate educational ambition of the programs should be the development of a national school curriculum to meet the 'inescapable duty and privilege for the older members of the community to teach girls and boys'.[1] But before this could be realised, the association had to develop, introduce and test standardised female-centric and led educational clinics, lectures and books at NBCA/FPA sites to disseminate sound scientific and medical advice and instruction to adults and prospective parents.

As I will demonstrate, following Wright's lead, between 1930 and 1950 the association became convinced that contraception and sex education and advice belonged within the formal education system and should ideally be provided there, and worked hard to broaden the reach and capacity of British sex and contraceptive education programs toward that end. Two distinctive teaching agendas became apparent through the association's efforts: both a formal clinic- and advice-based education for clients and medical students and practitioners; and an activist driven cooperation with other local and international marriage and family focused groups to check disparate and potentially contradictory lessons provided by ideologically diverse voluntary collectives. The association, in its dedication to standardising its methods and lessons within the bounds of verifiable science, worked to lead by example. To this end, many of its prominent female medical staff were emboldened to use their clinic experience to devise and deliver plain and accessible lessons for the public, and as younger generations of medical professionals became increasingly enamoured with contraception, to dictate and standardise a serviceable curriculum for teaching contraceptive theories and methods to interested doctors and nurses.

These increasingly shifted away from female-centric, personalised, in-clinic advice and agenda-tailored guides which continued to be provided in clinics, towards the association's stated goal of developing a state implemented and regulated, age-appropriate syllabus, geared towards providing adequate scientific, medical and moral advice. The association's work in this vein was galvanised by the 1947 Royal Commission on Population which publicised and ultimately supported the creation of standardised sex and contraceptive education for Britons. To capitalise on the Royal Commission, the NBCA/FPA actively sought and fostered (sometimes tense) relationships with other voluntary groups invested in sex and contraceptive education to work towards shared marriage, family and sex education goals. The last part of this chapter explains that much of the association's educational agenda was defined and promoted in partnership with other important contemporary lobby groups that took these subjects as their focus, namely the Eugenics Society, National Marriage Guidance Council, the Family Relations Group and the Central Council for Health Education. These groups loosely unified to promote their shared sex and contraceptive education and marital guidance agendas, whilst maintaining their autonomy to pursue specific goals. The NBCA/FPA actively undertook some prudent educational work, but typically provided medical and scientific support to the education programs devised and managed by these other groups, instead of managing them through the already physically and financially strained confines of its clinics. The NBCA/FPA and its partners utilised medical science, social support and government programs and initiatives to broaden access to teachings about sexual, marital and family planning matters.

The association, with Wright's assistance, identified three key contraceptive networks that needed to be cultivated in order to achieve scientific and practical legitimacy and acceptance of contraception: public and user networks, medical networks and contraception, marriage and family education and support networks. This chapter considers how each network was built and fostered using the promise to create and disseminate standardised sex and contraceptive knowledge as the key focus of shared interest and advocacy.

Defining the Wright Association Standards for Lay Sex and Contraceptive Education

Helena Wright's contribution to sexual and contraceptive education is renowned and has been well covered by Hera Cook, Barbara Evans and more recently Caroline Rusterholz, who have all focused on her role in the scientific and medical legitimisation and provision of sex and contraceptive education.[2] Wright dedicated her life to providing women effective and verified means to control their reproductive health and fertility through her work as a missionary, physician, contraceptive medicine specialist, educator and psychosexual therapist. From 1927, Wright helped to shape and direct the North Kensington Women's Welfare Clinic, and later NBCA/FPA and International Planned Parenthood Federation medical activities, and was an active campaigner for government funded and managed contraceptive services.[3] She spearheaded and sat on association executive and medical committees; and published leading educational manuals on sexology, female sexual fulfilment, fertility control, marital harmony and contraceptive technique for both medical and lay audiences. Her contribution to standardising sexual and contraceptive clinical practice and education cannot be overstated.

Wright was part of the second generation of British medical women. Early in the twentieth century she trained at the London School of Medicine for Women and the Royal Free Hospital and in 1914 became a member to the Royal College of Surgeons in England and Licentiate of the Royal College of Physicians.[4] In 1915 she received her MBBS (Bachelor of Medicine, Bachelor of Surgery) from the University of London, and trained at Hampstead General Hospital and the Hospital for Sick Children, Great Ormond Street where she worked under renowned surgeon, eugenist and contraceptive advocate, Arbuthnot Lane.[5] During the Great War, Wright worked at the Bethnal Green Hospital, treating military casualties. There she met her future husband Captain Peter Wright, a surgeon with the Royal Army Medical Corps and the couple wed before the war ended.[6]

This was the first life event that fundamentally altered Wright's medical focus and career. Before they wed, neither partner had any sexual experience, and this naiveté led to some sexual difficulty and maladjustment. In recalling her early sex life at the end of the twentieth century, Wright explained that she 'accepted what was happening to [her] and did [her] best with it'.[7] This highly educated and experienced medical pair had no knowledge of or access to contraceptives, and so when Wright almost immediately fell pregnant, she was forced to resign her medical post. This recollection likely exaggerates the couple's ignorance to make a broader political point that the lack of accessible and comprehensive sex and contraceptive education available in 1917 effected even her own comfortable life; and that her realisation of this gave purpose to her later life and pursuits. During her pregnancy Wright met Marie Stopes and the women became fast friends. Stopes, the wildly successful author of *Married Love* (1918) confided her plans to open Britain's first birth control clinic and her contention that women must be 'concerned in the physical side of marriage'.[8] Stopes invited an enthusiastic Wright to critique her follow-up *Wise Parenthood* (1919), which argued that sexual technique and contraception were wedded concepts. This exposure to scientific analysis of sexual and contraceptive matters for a general audience, in combination with her sexual and childbearing experience, likely motivated Wright's 1918 transition to gynaecology. In *Sex Fulfilment in Married Women* (1947), Wright claimed her interest in sex education began in 1928 *after* joining the NKWWC, however her transition to gynaecology a decade prior indicates the shift began earlier but had not yet formally extended to contraceptive and sex education.[9]

After Wright's gynaecological studies were complete, the couple moved to China to undertake missionary work. As an assistant gynaecologist at Shandong Christian University, Wright gained practical skills and experience dealing with health, psychological, educational and sexual concerns that shaped her career at the NKWWC, NBCA/FPA and later with the International Planned Parenthood Federation.[10] High mortality kept infant survival low, but the strains of repeated, closely-spaced pregnancy meant many women suffered 'crippling disabilities... such as prolapsed uterus, perineal tears, infertility, painful scarring and vesico-vaginal fistulas'.[11] These gynaecological ailments of unrestrained fertility and poverty required precise treatment and as Wright gained this expertise she determined that accessible education would render these conditions mostly preventable. Wright later applied this understanding in NBCA/FPA clinics to teach patients to prevent gynaecological ailments through physiological, sexual and contraceptive instruction.[12]

Upon returning to London, Wright established a private gynaecology practice, but was ignorant of the evolution of the contraceptive movement during her absence. She knew of available cap, condom and spermicidal contraceptives, but little of the church and medical profession's opposition to Stopes' publications and activities.[13] In 1927, there were now two major contraceptive clinics in London, Stopes' Mothers Clinic and the Walworth Women's Welfare Centre[14]; neither advertised they offered 'birth control', but it was recognised that mothering services were complimented with contraceptive instruction and sales.[15] Fierce opposition rallied at each site intimidating staff and prospective patients. 'Eggs and stones were thrown.... Onlookers shouted "Whores" and "Abortionists" [and] mocked clinic attendees and staff'.[16] Despite efforts in 1924 to solicit the Ministry of Health to offer contraceptive instruction at mother and infant welfare centres,[17] contraception remained absent in the medical curriculum and practice.[18] Medically experienced backing was needed to broaden contraceptive education: Wright and the association answered the call.

Wright reconnected with Stopes but was disappointed to find her paranoid and untrusting. Stopes was frustrated that her lack of medical credentials was cited by the profession to exclude her efforts, whilst simultaneously maintaining the non-medical nature of contraception. Doctors further opposed Stopes' vision for nurse-led medicalised services, decried the current lack of proof of contraceptive quality and efficacy, and warned of possible carcinogenic risks.[19] Wright, now with years of experience in gynaecology, agreed that medicalised contraception should be physician-led, and sought a better fit for her skills and aspirations. At the Society for the Provision of Birth Control Clinic's (SPBCC) Walworth site, Wright found a fully functioning clinic managed by an all-female medical and lay staff. This suited her self-serving belief that contraception should be a medical speciality, and all advice 'should be given by gynaecologists specially trained for the purpose, of whom there were none in existence'.[20] This view won her an offer of employment at the NKWWC as the Chief Medical Officer; in that capacity she helped to reshape its medical and scientific structures and standardised clinical practices and education. Public and user education became an immediate clinic focus.

Between 1930 and 1960, Wright worked to provide and disseminate accessible and correct sex, physical, physiological, psychological and contraceptive education. She wrote and revised 'simple books [and pamphlets] on birth control' to loan or sell to patients.[21] This chapter attempts to bolster the approach taken by Alison Bashford and Carolyn Strange in their study of sex education and mass communication, to use Wright's work and works not as mere 'evidence' of expertise, but to consider the medium, structure, contents and audiences of her messages as objects of enquiry.[22] As a sex educator, Wright focused on promoting contraceptive instruction as just one part of a holistic approach to marital, sexual and physical satisfaction. This aligned with current and future marital counselling trends that saw the emergence of sex therapy sessions for patients whose sexual and family aspirations remained unfulfilled despite generic medical guidance.[23]

Over the forty years of her career, Wright's tone and attitude towards sex and contraception evolved. As cultural shifts became increasingly tolerant of sexual topics and practices, she had abandoned her rigidly pure science style for a liberal social scientific framing. Wright progressively advocated towards a national standardised sex-positive education to destigmatise human sexuality and eradicate sexual anxieties. With these initial goals, Wright developed and implemented programs for the

NKWWC and association that aimed to standardise popular sex education, normalise and universalise contraceptive use, and liberate sex from outdated marriage and procreation-based socio-cultural constraints.

This last aim became a shared medical and political interest of the NBCA, and it worked to help Wright share her experience and engender goodwill to effect this agenda. In 1930 Wright was invited to lecture the annual Anglican Lambeth Conference's sub-committee on marriage and sex. There she presented sociological data on the negative relationship between poverty, health, sexual ignorance and birth control that she had obtained from her clinic experience to the three hundred Christian Bishops in attendance.[24] She exploited her identity as a 'relatively young [Christian] woman doctor' to highlight that her contraceptive work was primarily remedial and targeted virtuous Christian wives and mothers who, through ignorance, 'had more children than they could afford'. She explained that proper marital and sex instruction and education would essentially prevent these women's socio-medical woes.[25] After her speech, two resolutions were passed: that 'in all questions of marriage and sex the Conference emphasises the need of education', especially for those soon to marry, and that contraception be allowed.[26] The Bishops agreed abstinence was the 'primary and obvious method' but others were acceptable if used in line with 'Christian principles' and devoid 'of motives of selfishness, luxury, or mere convenience'.[27] These resolutions echoed then Archbishop of Canterbury Cosmo Gordon Lang's assertion that he would prefer 'all the risks which come from a free discussion of sex than the great risks run by a conspiracy of silence' on the topic. He advised that liberating sex from universal negativity would cure many current individual and social ills.[28]

The association's practice and data were central to creating an impression that the medical and religious communities' responsibilities to the public regarding sex and contraceptive education were the same: to allow and provide accessible scientific facts and instruction. In light of this support for sexual education, Wright published *The Sex Factor in Marriage* (1931), an applied scientific sex manual based on her NKWWC experience. To further employ the religious endorsement she had earned, Wright cunningly solicited a 'frank [introduction]... as a minister of religion' from the Reverend A. Herbert Gray and widely advertised her Christian and medical credentials to reach as broad an audience as possible. The book introduced sex with scientific accuracy and spiritual sensitivity and defined a minimum necessary sex comprehension for marriage that spanned anatomy, function, practice and morality. Her sex-positive approach directly opposed other current literature that offered a 'string of thou-shalt-nots'.[29] This tome set a minimum standard of sexual fluency for the British public and the association's clinics became central to its achievement.

The association and its medical committee were keen to implement this education as it was agreed that without proper sexual literacy, contraceptive education was also useless. Therefore, the lay educational efforts of the NBCA/FPA were actually implemented to ensure the efficacy of their central contraceptive programs and agendas. Here once again the female kinship that was called upon to encourage clinic patient attendance and trust was exploited to educate clientele about human bodies, sexual practices and reproduction. In providing a practical education Wright called upon her own progress from ignorance to enlightenment to counter the 'almost unbelievable level of public ignorance'.[30] She admitted that her own knowledge of sexual function came 'from textbooks' and intimate trial and error 'not from gynecologists who should have taught' these subjects.[31] Wright's religious convictions and sexual autodidactism definitely influenced her medical and pedagogical outlook and methods. In fact, the technique she, and by association the NBCA, advocated to counter sexual ignorance and encourage contraceptive proficiency deeply fused sexual science, spirituality and psychology to guide couples to sexual and reproductive satisfaction.

Wright's first text taught the physiology and biology of sex to remedy/prevent marital and coital disharmony and dissatisfaction.[32] Her second, asked and frankly answered the vital question *What is Sex?* (1932). It explained the evolutionary purpose of sex, the sex organs and their function, and the mechanics of 'the perfect sex act'.[33] This work placed human sexual procreation within the evolutionary canon of plant and animal reproductive biology where 'man stands at the head of the list as the highest animal'.[34] Sex, utilised by complex, multicellular beings, fused cells from each parent to form a genetically unique individual with discernible generational differences.[35] The sexual education offered in this text presented human reproduction as no more or less natural and scientific than that of an amoeba, earthworm, or dog. This matter-of-fact approach afforded sex and reproduction biological clarity as aspects of nature without religious, political or social baggage.[36] Using written and visual metaphors to describe human reproduction had a twofold benefit: it employed common language that was guaranteed to be easily understood, and demystified and destigmatised the subject matter. Wright actively equated human and animal sex organs, functions and structures using accessible semiotics: 'each testis is shaped rather like a chestnut; each sperm tube [a] hardish cord about as thick as a pencil; the penis has an acorn-shaped tip; the whole womb is shaped like a pear'.[37] The need for such simple and accessible examples demonstrates the lack of biological and physiological comprehension exhibited by contemporary patients and also illuminates the minimum sexual knowledge the NBCA deemed necessary. Mid-1930s, sex education became a core medical sub-committee focus as members pushed for formal 'Sex Instruction' to address ongoing ignorance and meet the demands of pre-married patients.[38] This scientific tract marked the association's first attempt to shape patient's sexual literacy and language.

Comprehensive explanatory images, diagrams, films, and hands-on activities were incorporated into NKWWCled sex and education efforts. One of the most important basics that Wright and the NBCA agreed was that 'general physiology', 'marriage and babies' and 'love, sex and conduct' were each different, yet complementary and intertwined, aspects of life.[39] So, it was vital to clearly illustrate and differentiate between 'the [body] parts that actually do the work and the parts that are only capable of sensation'. These illustrations, see images, supported Wright's apt and accessible description of reproductive systems as 'working factories'; parts where 'life giving cells' are produced, their routes of travel, and where they safely settle until use, and those 'special places on the outer surface of the body studded with special nerve endings'.[40] Establishing this disparity was fundamental, as in men the orgasmal and reproductive parts were 'inseparably linked'' and their systemic harmony obvious.[41] However, Wright took great care in rendering the female reproductive system to exhibit the compartmentalised nature of female sensory and reproductive functions to explain the latter was often not directly linked to sexual pleasure. As society became more accepting and open regarding romantic marital sex, the NBCA and Wright separately and in their joint ventures, produced books, educational lectures and clinic sessions that focused on female sexual function and disfunction. Male and female sexual satisfaction had long been associated with happiness, beauty and health, but colloquially sexual success was exclusively embodied by the male orgasm.[42] Thus, judged by this marker, female sexual success and failure, when it became a subject of study, was typically focused on the latter. In the *Sex Factor in Marriage* Wright mapped the female physiological sites 'with the capacity for responsive passion', tapping into the romantic sex ideals that had become the norm and were, in their ideal form, embodied in mutual orgasm. But Wright also placed the onus of responsibility for sexual success on women, who would 'certainly spoil [their] marriage' if they failed to become a sexual success.[43]

Wright widely prescribed the method she had used to overcome her own sexual frustration. The 'perfect sex act' she explained must stem from a strong *mutual* desire centred in the sex organs but which permeated each partner's entire body. This desire should result in instinctual rhythmic movement. As that movement quickened in pace, it should deepen to become an all-encompassing and mind and spirit altering feeling. At its peak the partners '*together* reach a climax ... marked by the sudden flowing forth of seminal fluid, and by a series of short, rapid muscular movements by the woman'.[44] This sex instruction taught that the act was not merely a mechanical, physiological processes, but 'a pleasure of the soul, attained for a few seconds, bringing with it a dazzling glimpse of the Unity which underlies all nature'—part scientific and part spiritual. This was a culturally apt sexual standard in 1930, however, within two decades, Wright observed and then embodied a cultural shift stressing psychological and emotional aspects of female sexuality and their links to marital harmony.[45] The complicated intersections of human consciousness and emotion made 'sexual union at once the simplest and most complicated act [the human body could] accomplish'.[46] This transition to a preemptive focus on female sexual failures was directly linked to NBCA clinic services through the pre-marital education program that aimed to prevent dissatisfaction by providing women with realistic sex and marriage education before they wed.[47] Wright admitted that her first educational efforts had caused some distress even among sexually literate couples, for whom her 'perfect sex act' prescription had failed.[48] This meant that an oversimplified and universalised sex education was insufficient to deal with the natural variety of female sexuality, psychology and inhibition. Wright 'bec[a]me dissatisfied and disappointed with [her teaching] results' and increasingly 'doubt[ed] the efficacy of the penis-vagina combination for producing [female] orgasms'.[49] During her two decades working in NBCA/FPA contraceptive and gynaecology clinics and psychosexual therapy sessions, Wright realised sex education and instruction must be idiosyncratic and subsequently promoted sexual practices to maximise the prospect of female orgasm.[50] She retracted her previous advice regarding attempting mutual climax on the grounds that her FPA clinic and her private gynaecological practice experience proved that the 'penis-vagina fixation in the mind of the public' stifled sexual learning and experimentation, and inclined against female pleasure because it avoided the epicentre of the female orgasm: the clitoris.[51] The doctor railed against this imposition of male sexuality upon women and in her private and association-related capacity articulated manual stimulation methods to 'awaken [women's] orgasm capacity', but cautioned that these suggestions were not comprehensive and couples must self-experiment.[52] Wright claimed that this more liberal and individualised instruction 'proved fruitful and constructive to a great many patients' but conceded that success was not guaranteed and so tenacity and malleability were key to sexual compatibility and success.[53]

Thankfully for Wright and the association's educational standing, her instruction on female sexual dysfunction became well-respected. This was in spite of Australian physician Norman Haire's condemnation of her claim that female orgasm capacity was near universal. Haire, the 'most prominent [interwar] sexologist in Britain', was outraged by this assertion, and encouraged readers to discard her libelous notions and book claiming that 'woman was never intended by Nature to have an orgasm at all'.[54] As Rusterholz has demonstrated, Haire's hostility to Wright resulted from their recent rivalry to define the ideal method to test the safety and efficacy of the Gräfenberg ring. Wright's affiliation with the Birth Control Investigation Committee (BCIC) and NKWWC won approval for her scientific methodology from the local birth control community, and Haire was sidelined for lacking scientific rigour and providing inadequate evidence.[55] Despite this detractor, Lesley Hall, Hera Cook and Kate Fisher have lauded Wright's radical sex-positive brand of sex education.[56]

Sexual compatibility and marital harmony, female sexual awakening and success and preparation for marriage became central NBCA goals mid-century as the organisation worked to effectively shirk the growing negative public conception of the term 'birth control' and the reality that association clinics provided more extensive holistic medical and educational services. Thus, it is no coincidence that the practical, educational and attitudinal skills the association prized and worked to export were also those required for contraceptive success. Patient persistence, open-mindedness and tractability were also being standardised across the lay sex and contraceptive education networks that the NBCA serviced.

But in individualist and private matters like sex and contraception, dedicated instruction was no promise of individual success. That demanded a comparable commitment to learning. Glimmers of desire existed amongst association patients post-war when Wright began to openly advocate for 'education and preparation in sex matters [to be] given to young people from childhood onwards'.[57] Contraception, however, was entirely omitted from all of Wright's sexological works.[58] Sex and contraceptive education were seen to be vitally connected by contemporary contraceptive lobbyists like the NBCA/FPA; but the impropriety of openly promoting the practice even as a cure to psychological impediments to sexual fulfilment 'such as stress ... caused by an unwanted pregnancy, or the fear of one', was appreciated by Wright, even in her capacity as one of Britain's leading sex therapists and contraceptive physicians and activists.[59] This omission reflected the fact that sex and reproduction were increasingly being considered worthy objects of scientific study, whilst contraception was not. In order to increase her sex education reach, Wright divorced these subjects, and exclusively created contraceptive training materials in line with association demand and methodology.

Wright's Lay Prescription for a Healthy Family and Sex Life: Birth Control

In her career, Wright published two tracts solely dedicated to contraception, one for a lay audience and the other for the medical profession. Mid-century sexual and contraceptive education commanded mass-attention from birth control advocates, doctors, eugenists, voluntary societies and government health and education bodies—but many were taciturn to begin popular education without scientifically verified lessons and methods.[60] In contrast, the NBCA/FPA was tightly focused on advertising and facilitating 'birth regulation' and family spacing knowledge through books, flyers, lectures and word-of-mouth. With its expansion, the association developed education and instruction lectures and programs to advertise its officially verified contraceptive methods.[61] This educational agenda relied on local health authorities and voluntary societies establishing visible contraceptive clinics and progressively injecting contraceptive prescription and instruction into broader social, medical and economic debates.[62]

With increased publicity came amplified clinic and lecture attendance, press attention and funds.[63] This overt exhibition of public interest and support for contraceptive instruction led the NBCA medical sub-committee, under Wright's chairmanship, to establish a 'comprehensive library on birth control covering both the national and international field'; this was available for loan to branches and members to improve and standardise association-wide instruction techniques. In October 1934, the medical sub-committee criticised the current dearth of concise lay literature and resolved to 'issue [and oversee] authoritative information dealing with every aspect of' birth control.[64] It also undertook to counter authors and publishers peddling 'unsatisfactory and inaccurate' information with medical evidence and to request 'withdrawal' or revision of all such texts it identified.[65] It further offered, and was allowed, to review contraceptive literature for the medical press, specifically, the *Lancet* and *British Medical Journal*[66]; finally it published and distributed clear instructions when regulatory efforts failed.

The association's prime interwar successes in this arena came from circulating accessible contraceptive advice for patients unable to access contraceptive clinics through its annually revised brochure *Advice on Family Spacing*.[67] Finally, its medical affiliates, Carlos P. Blacker, Joan Malleson, Margaret Jackson, Gladys Cox and Wright, each strove to publish and present clear and tested advice and instructions for the masses.[68]

Wright was particularly well placed to devise and disseminate comprehensive medical and scientifically verified contraceptive advice, due to her appointments to the NBCA executive, medical sub- and ad hoc sub-committees, and her work with the press and publicity sub-committee.[69] Her publication, *Birth Control: Advice on Family Spacing and Healthy Sex Life* (1935), applied her vast professional experience to describe practical, scientific and moral reasons supporting contraception and argued for a uniform national contraceptive instruction program. This book attempted to counter anti-contraceptive prejudice with clear and reliable scientific advice to allay moral and methodological 'confusion in the public mind'.[70]

In the mid-1930s the term 'birth control' was almost as contentious as the practice,[71] but Wright explained the 'real object' of the modern contraceptive movement was the creation and protection of 'happy homes peopled by happy children'. She and the NBCA aimed to promote 'family spacing' to ensure every British child was wanted and planned. Thus, parents could reproduce at 'intervals ... that allow mothers [to] be and ke[ep] in the best possible health' in order to offer all children their best health and security prospects. Wright employed the still in vogue eugenic principles of heredity and environment to warn that 'bad stock' which bred wantonly most needed contraception to ensure no families were 'ruined' by producing 'more children than' could be raised in a 'minimum of decency and comfort'. Contraception was thus an 'inevitable step in the advancement of civilization'.[72]

Birth Control cited medical and scientific evidence to counter common objections that 'unnatural' mechanical methods caused harm, sterility, or moral and physical threats to families, marriages and health. Statistics proved maternal mortality spiked following multiple pregnancies and confirmed children born into large families typically suffered poorer overall health.[73] Sterility and harm were disproved by the BCIC's scientific research and further assured by the NBCA's approved list of clinic-supplied products (see Chapter 4).[74] Contraception was also claimed to cause affairs and sexual promiscuity; but Wright assured there was no link.[75] Contraceptive technologies were compared to contemporary technologies of health and convenience such as 'tooth-brushes, umbrellas, or disinfectants' which were 'invented to fit a special purpose of necessity or convenience' and she promised that each association approved method could be employed just as confidently. A 'reliable method' must 'effectively prevent conception' without interfering with sex, 'be entirely harmless; ... easy to learn; and cheap to buy'.[76] In 1935 no one method satisfied each criterion. So, Wright echoed the association's dual-technique that combined 'a rubber appliance to protect the mouth of the womb' with a 'chemical preparation capable of destroying... sperm'.[77]

The association endorsed two varieties of rubber contraceptives: diaphragms or caps that women inserted prior to anticipated sex and required prescription and fitting by a physician; and condoms or sheaths, which were slid over the erect penis immediately preceding sex. All were made from quality rubber, priced reasonably and easily obtained. All washable rubber contraceptives required proper storage and upkeep for long-term efficacy. These prophylactics were to be paired with 'sperm destroying chemical preparations'. Wright advised the only guaranteed spermicides were those tested and approved by the NBCA.[78] An interesting assertion, as Chapter 4 and Claire Jones' work on the contraceptive trade both demonstrate is that the association's standards were somewhat malleable and occasionally failures where accepted in cases where a long-term relationship existed between the NBCA and manufacturers.[79]

As Bashford and Strange argue, successful sex education demands trust between the educator and pupil, the former must provide 'reliable information' and the latter must accept this advice has a base in clinical authority.[80] Hence, Wright's emphasis on the NBCA and BCIC's scientific testing and evidence of the safety and efficiency of her recommended methods. Her manual also conceded the existence of informal lay contraceptive knowledge networks that presented risks and ramifications for unwary users. In order to rely on expertise and experience, Wright exclusively advocated contraceptives that the association had verified were adequate for purpose. This credit allowed readers to grasp Wright's altruism in creating such an accessible and trustworthy guide for women she was unlikely to ever meet.

Birth Control also featured explicit instructions for proper application and use of each method, along with hand-drawn diagrams depicting perfect and improper fit. For remote readers, she described a simple and cheap method that offered a 'reasonable hope of success': cotton wool soaked in vinegar solution. The wool must be fashioned into a ball (sized to fit the user) tied with a long thread for removal. Immediately preceding use, the ball should be dipped into an equal vinegar/water solution, squeezed and then inserted deep into the vagina. This was a poor and distant woman's version of the association's most favoured cap and chemical routine. That method was practically the same except a proprietary spermicide and lubricant were applied to the cap which was to be inserted per the physician's direction. For satisfactory protection from potentially dormant sperm, it was an 'absolute rule' that all insertable contraceptives 'must be left undisturbed for ten hours after intercourse'. Removal of the sponge and cap were identical except the former was discarded, and latter 'removed, washed, dried and [returned to] its box'.[81]

Wright assessed and presented the merits of all available methods for 'safety, aesthetic comfort, and practical convenience'. She reported that male barrier contraceptives could easily be tested by blowing air, smoke or pouring water into sheaths and condoms to reveal holes, but diaphragms and caps could not. Insertable methods (when used properly) were near infallible, offered superior comfort and caused no 'blunt[ing of] the man's sensations'.[82] Women's pleasure and sensation, which formed such an integral part of her sexological tomes, was entirely ignored here. Obviously, there was some sense that these were distinct audiences that demanded independent content. Wright claimed there was nothing complex or frightening about contraception to stop capable couples consistently applying two methods, but recommended routine clinic visits for absolute security.

Birth Control concluded with a practical plea to support a standard 'organised instruction' for Britain. The call marked Wright's book as unique in comparison with contemporary tracts as it stressed the 'urgent social necessity' to teach every married woman safe and effective contraceptive techniques.[83] Pragmatically, Wright conceded the service would initially be the responsibility of the association but argued that this collective had already and would continue to 'set up… clinics to instruct mothers to whom further pregnancy would be detrimental'.[84] A wider, more formal education should be demanded by those most impacted by its omission from school and social education. She reminded her frustrated audience that local health authorities and physicians had so far failed to provide contraception even according to Ministry of Health provisions and insinuated that this was a dereliction of their duty of care. Wright as a practicing physician clarified that modern tools 'offer striking and immediate good results' and that official delays providing the service would only be remedied if loud community agitation demanded the service. She confided that contraception could 'be safely taken over by the government' as it clearly fell within the purview of the public health service and should not be left to the 'responsibility of the generous few'.[85] She linked public education and demand directly to the medical profession's reticence towards scientific and medically sanctioned contraception claiming; 'as soon as [doctors] understand their patients are serious in their desire for contraceptive instruction, the doctors will obtain the necessary special training'.[86]

In this publication Wright exploited her association-based experiences with contraceptive medicine, science and education to push the public to demand contraceptive education and access. By enticing readers to desire a government-led standardised service, Wright tapped into public support for contraception as a public health service. This was concurrently attracting an emerging generation of medical practitioners, educators and politicians. Support for contraception was not quickly or easily transplanted onto the medical profession, however, this agitation definitely translated into a broader public knowledge network that demanded medical contraception and aided the association's concurrent foray into contraceptive training for doctors.

The National Birth Control Association and Contraceptive Training for Doctors and Nurses

The association and its leadership including Wright were explicit about the existence of two distinct audiences for contraceptive training and education: public and medical knowledge networks. The latter, on the whole, proved particularly unreceptive and 'strangely discordant and ill informed' on the issue until mid-twentieth century. In 1934, NBCA medical staff complained there was 'no instruction on contraceptive methods' at medical colleges despite the Ministry of Health recognising contraception as a medical therapy.[87] This omission prompted the association to encourage some prominent female medical staff to devise, publish and implement medical training manuals, lectures and later standardised medical instruction courses for medical students at the Walworth (WWWC) and North Kensington Women's Welfare Clinics.

In 1922, as a Malthusian League clinic, the Walworth site instituted lessons in contraceptive theory and technique for interested medical students and staff. Its Medical Officer Norman Haire intermittently presented sparsely attended lectures on 'sexual anatomy, hygiene and birth control' for physicians and held practical application sessions immediately after.[88] Rusteholz correctly identified the prime reason for the unpopularity of these early session as being fundamentally tied to the generational mores of older and emerging physicians. Women doctors were the most likely to be interested in obtaining contraceptive expertise as this was emerging as a new branch of women's medicine and many personally experienced the result of a lack of this knowledge in their own marital and family lives.[89] It was to serve and service this generation that medical training expanded significantly after 1924 when the SPBCC took over the Walworth site and founded the North Kensington centre. The society intended to form new clinics around the country and this goal demanded trained medical practitioners and qualified nurses. So, formal lectures and training sessions were quickly instituted at both sites. Under the direction of superintendent Evelyn Fuller, Walworth liaised with several London teaching hospitals to propose training medical staff and students interested in contraception.[90] Despite some enthusiasm, before 1930 most doctors were averse to the practice, except under specific medical and ethical caveats.[91]

In 1922 the National Council of Public Morals in cooperation with the National Birth-rate Commission studied medical aspects of contraception. The WWWC's Chief Medical Officer Gladys Cox submitted evidence that her clinical experience at Walworth proved most patients required contraceptives for health reasons, rather than demonstrating 'improper' social or economic motives.[92] She described the average clinic patient as a mother with an infant at her breast and a toddler tugging her skirt, who had been sent by a friend or neighbour, or else referred by a hospital or general practitioner. 'The majority of [her] patients' had 'lacerated cervices' resulting from childbirth and thus contraception was therapeutic. Cox described typical WWWC clinical practice. Each patient received 'a gynaecological examination by a doctor, who also fits the pessary'. Guidance on pessary 'manipulation was [then] offered by a Clinic Nurse' and half-annual reviews were recommended for follow up 'gynaecological examination by a doctor, and to obtain a new pessary'. Patients unsuitable for fitting were 'referred to [a] hospital or private doctor for preliminary treatment'.[93] Cox's evidence, based on the analysis of 7,600 SPBCC patients treated between 1920 and 1926, was intended to appeal to physicians who in ignoring contraceptive responsibility were breaching their Hippocratic oath to ensure their work did no harm. Former President of the American Gynaecological Society and contraceptive advocate Robert L. Dickinson bolstered these findings with evidence from his own practice. He travelled to Britain to represent the United States Committee for Maternal Health that regarded his contraceptive research and experience 'most likely to be of scientific value' in evincing its perspective that 'physicians alone could adequately supervise contraceptive' matters.[94] Dickinson tied the need for contraception to progress that had been achieved by active medical and scientific interventions on bodies and in public spaces to improve public health, fertility and life expectancy for almost two centuries.

> He [the doctor] has taken his part in fostering [the] increase in population through successful contest with preventable disease and infant mortality and … [is] now expected to do his share toward solving the problem.

> The two choices … are: to investigate medical aspects of control and conception, or to decline to investigate.

He warned, the medical profession's exclusion of contraception from textbooks, university lectures, discussion at representative meetings and 'authorized *clinical* investigation', would inevitably prompt a group without training or experience to undertake these functions with a non-medical bias.[95] The commissioners agreed, and resolved that clinics should be guided and controlled by 'experienced and judicious medical practitioners' and 'hospitals should [provide] advice and instruction on the subject'.[96] This acknowledgement from the profession marked the formal origins of the medical knowledge network that the association subsequently worked to inform and later, standardise.

This unexpected recommendation received prompt but contrary attention in the medical press. The *British Medical Journal* downplayed the concession explaining that this 'interim' report had 'not set the seal of finality on the contentious subject'[97]; whereas the *Lancet* and *Public Health* faithfully conveyed the committee's assertion that the profession should cooperate with clinics and provide contraceptive advice.[98] No medical publication was yet willing to advocate contraception, but the prospect of contraception being a subject for inclusion in the medical curriculum had emerged. Eugenist and founding member of the BCIC, Carlos P. Blacker dubbed this 'an authoritative medical pronouncement' and predicted that the current cohort of young doctors and students would eagerly embrace contraception as there was evidence of growing numbers that agreed with Dickinson's position.[99] Physician and educator Roger N. Goodman added that young physicians had no doubt about the usefulness and appeal of contraception and were in fact approving and keen to engage.[100] This shift was confirmed by increased student attendance at early 1930s NBCA training sessions to gain theoretical and practical knowledge from experts in the field.[101]

In 1933 the NBCA determined to meet this new-found demand for medical training and formed a Training Sub-Committee to 'centralise' and 'cover the field thoroughly'. It determined to standardise training at the NKWWC and WWWC to facilitate the timely expansion of accessible medical contraception 'in the provinces'.[102]

The association's initial training method comprised 'one lecture attended by a crowd' followed by successive demonstrations for small groups; the Walworth method. This plan, however, was envisaged without considering disparate extant training programs, such as the 'demonstrations by doctors' hosted at Marie Stopes' Society for Constructive Birth Control Clinics. A medical training memo written after the committee undertook this survey revealed the association had to choose between two unfavourable options to achieve its medical training ambitions: 'either include [Stopes'] clinic in [their] list of those providing training; or omit it and prepare for a storm'. The memo reasoned for its exclusion as Stopes' favoured barrier, the 'Prorace' cap, had been deemed 'undesirable' by the NBCA medical committee; and further indicated that if her teaching, like her practice, was managed by nurses, this would provide sufficient 'ground for rejection', as only doctors, with sufficient gynaecological and contraceptive training were fit to prescribe, let alone teach the technique. This was a 'ticklish question' because Stopes was notoriously unwilling to compromise or cooperate with the association, even to achieve shared objectives.[103] The association had bitter experience with Stopes. In 1931 she was nominated to its executive committee, and shortly after members' fears that she would 'try to dominate' meetings were substantiated. Wright had initially refused to participate without Stopes inclusion and promised to 'manage' her friend. However, this proved impossible, and Stopes resigned in frustration.[104] Thereafter, the association sought to disassociate its medical training from Stopes'. The training committee scrutinised her methods hoping to repudiate her methodology on the grounds that instruction was 'done by nurses'[105] The association was determined to establish standardised, verifiable contraceptive instruction techniques produced by and for physicians to persuade doctors that this was a medical specialty equal to any other on the medical curriculum; and that it was being unfairly excluded for wholly unscientific, socio-cultural reasons.[106]

In February 1933, the training sub-committee announced an NBCA-approved medical training scheme for North Kensington and Walworth. It 'combined [a] lecture and demonstration for … 8–12 students' and would charge each five shillings. Training doctors could negotiate their fees and patient demonstrators were paid as fitting models. The NBCA promoted these sessions to the medical press, hospitals and medical school gynaecology departments.[107] The sub-committee proposed to divide hospitals and schools between the two centres 'according to their geographical position'.[108] However, the WWWC baulked as in its capacity as an independent SPBCC clinic it had already effected training arrangements with some schools and hospitals. It refused to relinquish these to the NBCA. Terse correspondence from WWWC superintendent Fuller advised the association committee 'that Walworth would continue to make its own arrangements about training' as its sessions were already running 'very satisfactorily' at 'full capacity'.[109] The association secretary Margaret Pyke countered that there must be 'some misunderstanding' on Fuller's end, as at the June 1932 medical training meeting representatives had agreed that the not yet fully integrated SPBCC and NBCA 'should work well together'; took that as an agreement to cooperate.[110] Actually 'the SPBCC [had] stated ... [it] had no objection' to the NBCA maintaining voluntary clinics and that it was 'willing to train doctors and nurses, etc'.[111] It had not agreed to have its independent work subordinated by administrators at a newly formed, and as yet unproven, voluntary society, and therefore strongly protested the NBCA soliciting or interfering with its own well established working relationships with hospitals and schools. In conciliation, the association agreed it 'would not interfere with any established connection between a clinic and a hospital'.[112] This tact ultimately ensured no overlap between the Walworth and North Kensington medical training agendas and paid reverence to the SPBCC's proven method by essentially taking it as the association model. Not all medical or even birth control advocate networks in the early years of medicalised contraception were universally cooperative. As I have demonstrated there was a very fine line between cooperation and competition even when groups and individuals definitely shared objectives, approaches and techniques.

The NBCA advertised its medical training programs at hospitals and universities with two distinct goals: to inform students and graduates of the resource, and to gauge and locate professional interest in and approval for contraception. Understanding current professional sentiment would allow the association to tailor its program to address both socio-political and technical aspects of the practice. Association President Lord Thomas Horder was its only senior enough member to circulate the appeal and in 1933 he settled to lobby at least five institutions.[113] The next year he petitioned significantly more hospitals and colleges.[114]

Per its agreement with the SPBCC, the NKWWC training program was directed by the association. This necessitated close communication between the NBCA and North Kensington medical committees to ensure its success. The arrangement allowed Walworth to keep servicing trainees from the five hospitals in its historic catchment and allowed the NBCA to direct students from all others to North Kensington.[115] Wright was then the NKWWC's Chief Medical Officer and chairman of both the NBCA medical sub-committee and its local equivalent; she pioneered and led the program. Wright also consented to be named in NBCA advertising, that advised administrators that the revered medical figure and contraceptive expert would attend the various colleges to 'give … a lecture to the students'.[116]

Horder's support and Wright's reputation proved inadequate to gain approval from university and hospital directors. No response was hostile or dismissive, and in fact each institute promised the submission would go before its medical curriculum committee for formal consideration, but each communication made clear that contraception had no place on the current program.[117] Any cooperation with the NBCA's medical training was also vetoed, as no professional institution would 'arrange systematic courses of instruction with non-medical bodies'.[118] This stance was maintained throughout the 1930s, the only concession gained was that institutions slowly undertook to inform students that NBCA training was available. So ultimately, the NBCA was forced to bypass educational institutions and focus its advertisements in the medical press.[119]

The immediate response was overwhelmingly positive. Within a year of the NBCA advertising its medical training service in the *British Medical Journal* and the *Lancet*, 'about 250 doctors ha[d] written for information' and tuition.[120] Fifty responses were received in just two months.[121] As the only named clinic, the NKWWC was inundated with applications and, to meet demand, amplified its increasingly 'popular training sessions'. In 1935 the association declared that even without official support the service was in high demand and undertook to 'establish some criterion of efficiency for doctors trained in birth control methods at Clinics'. To this end, the training sub-committee resolved to 'insist' on a minimum level of mandatory clinical contraceptive experience before a doctor be allowed to practice unsupervised.[122] This marked the NBCA's first effort to dictate medical contraceptive practice and define a standard of certified competence for accrediting contraceptive physicians. Although this ultimately took decades, after 1935 it became a primary training committee objective.

Initial qualification plans were defined that year. These proposed that, at least initially, 'adequate training' should require a student read Gladys Cox's *Clinical Contraception* (1933) or Joan Malleson's *The Principles of Contraception* (1935) or attend one 'lecture or demonstration', and that they must actively participate at 'three or four clinic sessions'. Although no certificate could be offered, the committee advised that training clinics could 'give testimonials' of proficiency.[123] Fifteen years later, this unofficial recognition was formalised with a register listing the training activities and results of association-trained contraceptive physicians.[124]

In mid-1930s, several promising developments occurred which led the association to believe hospitals and universities might shortly incorporate this work. In late 1934 the University of Cambridge Bachelor of Medicine exam included an obstetrics and gynaecology question regarding the medical necessity to avoid pregnancy and extant appropriate contraceptive methods.[125] Shortly afterward, University College Hospital initiated an outpatient contraceptive clinic.[126] Then in 1936 Wright, as an expert in contraceptive practice and technique, was invited to lecture the British Postgraduate Medical School on the subject.[127] Finally her medical peers had recognised her expertise and she presented the first contraceptive training course within a British medical institution based on the NKWWC's tried and tested medical training curriculum.

Under Wright's tutelage and supervision, North Kensington became a renowned medical training centre. Before 1934 its medical staff, mainly Wright, offered only medical lectures. After that demand so increased that the medical committee was obliged to reorganise training. Some staff proposed to allow doctors to learn how to manage, employ and apply available contraceptive techniques during birth control sessions. This was contentious as hosting and training students in real time 'might be very delaying' to both patient and clinician. Consequently, 'only experienced students' were sanctioned to 'see how the clinic is run', and separate introductory classes were held to teach the basics.[128] This interactive, gradated scheme was supported by the NBCA training committee as an appropriate structure for all training clinics and remained relatively unchanged until the 1940s.[129] In that decade every aspect of medical training was up for debate: qualifications, curriculum, the relationship of midwives' to contraception, minimum experience needed for proficiency, and sourcing and employing appropriate teachers for available methods.[130] The medical education developed and standardised at North Kensington under Wright's direction was also actively exported to other clinics.

In 1949 the FPA developed an interest in defining formal medical training guidelines. The NKWWC medical committee suggested that all association medical training should replicate its own methods, and proposed two radical new regulatory conditions be introduced. That

1. No FPA Certificate [was] to be issued to candidates without previous gynaecological experience;
2. Each candidate [must] pass through the hands of at least two [staff] doctors.[131]

In large clinics, like North Kensington, these recommendations were easily implemented. It still served the majority of London's medical students and retained a large staff to deliver and oversee training and certify competence.[132] But as new training clinics were approved and established, many in remote cities and towns, the possibility of maintaining NKWWC-defined standards diminished significantly. This prompted the FPA executive committee to move to centralise all in-clinic training and for teaching and accreditation records to be held and maintained at head office. Pyke relayed the committees belief that 'it would be very valuable to have a record of any doctors or nurses trained' and have all training clinics 'report to us when … training is completed, whatever the [student's] results'.[133] Initially this proposal was limited to graduates being considered for FPA clinic appointments, but was soon extended as 'hav[ing] a record of all doctors trained' would be 'absolutely invaluable' to the association's aspirations to set and guide national medical, practical and educational networks of contraception.[134] A *Draft Memo to Training Clinics* was circulated to clarify this was a necessary move as there were many doctors 'who either completed their training or worked at [North Kensington] clinic sessions' whose 'training was unrecorded'; and so, according to these new standards they would 'be expected to [re]take the training'.[135] The previously lax record keeping regarding training meant the association could not even guarantee the proficiency of currently employed clinic doctors. This was an outrage to its stated aspirations to control, direct and standardise contraceptive knowledge and practice. The ad hoc training committee, which Wright managed during the 1940s and 1950s, ruled that centralised administration would 'lessen the amount of record-keeping required' by allowing each 'active training clinic', six in 1949 expanding to 14 in 1950,[136] to submit to the head office a single monthly 'Training Report' listing each doctor and nurse currently undertaking the FPA Certificate, and their progress.[137] By the 1940s the FPA's dedication to standardising and regulating its medical training program firmly intersected with its contraceptive endorsement activities. The result was the creation and maintenance of an up-to-date register of association-trained contraceptive practitioners. Much like the manufacturers whose products populated the approved list (discussed in Chapter 4), enrolled physicians were those who perceived a professional value in gaining contraceptive experience for their current or future practice. Nevertheless, with the advent of a centralised system of record management and certification, a small but growing number of physicians facilitated a subtle transmission of educational authority to the FPA by subscribing to its approved contraceptive methods and techniques.

Through centralisation, the association was also able to mandate the gynaecological experience and proficiency trainees must possess to be admitted into FPA training and obtain certification. This decision was made despite NKWWC doctor's claiming it was unnecessarily restrictive and possibly detrimental to professional engagement with contraception. Forging ahead despite criticisms, the medical committee emphasised that all doctors could receive contraceptive training, but only those who met the FPA's competency criteria would be certified.[138] An adept student must be

> (i) able to do a rapid and accurate pelvic examination, (ii) able to spot gynaecological abnormality, (iii) [be] familiar with the use of a vaginal specula … [and] familiar with three basic types of cap and the mode of action and times of the various chemicals used.[139]

After 1950 more clinics around Britain expressed interest in offering training. In response the association defined an FPA-wide syllabus and fixed standards for the proficiency and experience of the increased number of instructors this would necessitate. By the 1950s the association took itself, and its medical, educational and propaganda agendas and standards very seriously. As will be discussed in Chapter 4, its success with the control and standardisation of contraceptive products and technologies via the approved list, provoked an enthusiasm to establish similar minimum standards of knowledge, practical proficiency, and methodology for the public and medical professionals. To the latter end, prospective training clinics and all their teaching staff were required to receive instruction and approval from an experienced FPA supervisor. Wright, as one of the association's most practiced educators, joined its newly formed inspection panel and carried out nation-wide reviews.[140] She also continued to oversee NKWWC training sessions to certify that all accredited graduates had obtained the required minimum hands-on experience; by attending '100 contraceptive sessions'.[141]

In 1950 the FPA released its first official training standard which defined its medical curriculum, and set minimum requirements, qualifications and experience for teachers and students, clinic structure and facilities, the fees payable and conditions for certifying doctors and nurses.[142] The articulation and enforcement of these standards was now feasible due to increased public, political and medical tolerance of and interaction with contraception; as well as the 1947 *Royal Commission on Population Report* which stated that contraceptive advice should be afforded to married couples 'as a duty of the National Health Service' and that 'the initial duty' for contraceptive advice and instruction 'should rest with the family doctor'. The report further recognised that ongoing professional hesitancy toward contraception was because doctors were 'not trained for the purpose', rather than there being a lack of interest or limited demand. Thus, the Commission broadly supported universal medical training and found this knowledge gap, that the FPA had long worked to fill, could easily 'be overcome by an adjustment of [formal medical] training' to incorporate the contraceptive content.[143]

Teaching the Wright Contraceptive Technique to Doctors

Noting a growing transition in support for contraceptive medical training, Wright perceived there was yet no 'practical handbook, full and detailed enough to provide a [medical] reader with all the information necessary to begin fitting and instructing patients'. So, she produced *Contraceptive Technique: A Handbook for Medical Practitioners and Senior Students* (1950) to reveal the 'principles and practices [she had] found effective' to doctors and nurses who believed that contraceptive instruction could solve the 'population problems of the world'.[144]

Wright's tract targeted physicians already sympathetic to contraception but framed its arguments as entirely medical. She introduced the practice of medically facilitated family limitation as an inherent duty of the physician by allying it to contemporary research from prominent physicians including Guy's Hospital Senior Obstetrician Harold Chapple who warned that 'uncontrolled maternity can become a fatal disease'. She further stipulated that planned parenthood was now a 'basic necessit[y] for the maintenance of reasonable conditions of life' and so, withholding effective techniques could be, at the very least, considered negligent.[145] It was essentially being argued that universal access to effective contraception enabled all couples to shape and time their family based on their financial and physical conditions, and that this sense of responsibility and control made 'children more desirable rather than less'. So, the medical practitioner with contraceptive knowledge was able to actively contribute to the physical and psychological wellbeing of Britons. There was indeed 'no more useful service medicine [could] do for society than to spread the policy of planned parenthood'.[146]

After refuting the profession's historical objections to contraception, the manual defined 'acceptable contraceptive methods'. Here Wright discernibly allied the guide to the FPA's standards by citing the dual contraceptive prescription as the safest for absolute surety; she justified the association's methods and preferred products and manufacturers by explaining that each had been proven scientifically sound based on laboratory and statistical tests that were reviewed and revised annually. In doing this, Wright actively attempted to invite physicians using this as a reference to accept and join the FPA's contraceptive knowledge network in order to universalise and share its findings about products and methods that could 'be used with confidence for the [contraceptive] purpose'.[147] For the uninitiated Wright took great pain to detail the various methods, their correct application and how to determine and fit a suitable cap for individual patients. She particularly explained the limitations of each method and product for patients with specific physiological conditions, and warned against 'generalised statements as to which type of cap is suitable for individual patients'—essentially, she was citing her own experience that proved prescription had to be tailored to each patient in order to preemptively argue against the practice of overreliance on a single method or brand. Jones has shown that many patients and birth controllers like Stopes and Haire did this with their favoured caps and manufacturers, often to the detriment of efficacy.[148] Finally, Wright offered practical lay teaching methods she had found effective. She explained that patients needed the 'clearest possible instruction' and described and illustrated practical methods gleaned from her FPA practice to teach and guarantee cap insertion, positioning and removal proficiency.

Importantly and uniquely, Wright included a comprehensive discussion of 'unreliable contraceptive methods' in the tome. She advised that association-led scientific and sociological research had officially designated several traditional, technical and mechanical methods as unsuitable, unreliable or dangerous: coitus interruptus and reservatus, the 'safe period', douching, using a condom/sheath, cap or chemical in isolation, as well as two intra-uterine devices, the Wishbone Pessary and the Gräfenberg ring.[149] None of these methods, barring withdrawal and the safe period, were suitable for general advertising or discussion, but physicians must be aware of their method, failure rates and health risks to provide wise and considered counsel and potentially prescribe these methods if and when they were best suited to individual patient's requirements.

In concluding, Wright clarified that the FPA's ambition to centralise and standardise British contraceptive practices, policies and education was ultimately so that most practical and political groundwork would be done for the Ministry of Health or medical profession when each inevitably assumed this work in perpetuity. Until now, in spite of the association's successes regulating contraceptive products and clinic practices, there remained a fundamental disconnect between it, the medical profession and government regarding contraceptive responsibility. The association had laboured incessantly since 1930 to prove that contraception was at its core a medical practice, yet the 'attitude of the Ministry of Health towards family planning was still unsatisfactory' as it still limited access to contraceptive advice, and failed to subsume the practice into the new National Health Service.[150] This was then, part manual and part appeal to doctors to alleviate the association's burden by providing the service wherever possible.[151]

The FPA training and medical committees deemed Wright's book so expert and comprehensive that medical practitioners/students who had read it were exempt from attending the introductory training lecture prior to practical instruction. The association later suggested that all potential trainees 'should read a modern book on Birth Control methods (preferably "Contraceptive Technique" by Dr. Helena Wright)' before commencing.[152] The work represented the fundamentals of the association's medical knowledge and vastly extended the capacity of its networking reach as it could serve as a reference for introduction, revision or in practice. Wright had produced a text that both complemented and guaranteed the educational supremacy of the FPA's medical training, by ensuring the continuity of its clinical methodology and exponentially spreading its technique. Essentially, the FPA medical training strategy and this book both ensured doctors and nurses received sufficient information and experience with contraceptive methods and matters to guarantee they were properly prepared to assess and administer products and advice in practice, whether they had access to in-clinic training or not. Despite not being able to personally guarantee all the contraceptive practitioners in Britain via its training register, Wright and the association remained dedicated to verifying doctors' contraceptive technique through direct instruction and observation in practical application wherever possible.

Wright continued to work as the prime medical instructor at the NKWWC into the 1950s when she developed contraceptive educational programs for foreign and travelling doctors and nurses through her work with the International Planned Parenthood Federation.[153] She further designed and led an intensive training course at the Marie Stopes Memorial Centre and offered comprehensive contraceptive training for the Royal College of Nurses.[154] One final method of instruction that Wright designed and attempted to patent was an anatomically correct 'life size, flesh-coloured plastic model of the female pelvis and upper thighs, complete with vaginal opening', which in addition to living models, was used for practical demonstration, experimentation and assessment in sessions. Ultimately, Wright chiefly considered herself an educator and spent her later years dedicated to this pursuit.

Wright and the FPA's medical training agendas embodied the educational subterfuge that the 1926 National Birth-rate Commission warned would emerge if the medical profession remained aloof from contraception. But fortunately, the FPA was not staffed by quacks or 'irresponsible persons', but medically and scientifically trained and socially responsible volunteers who built and applied their collective knowledge and experience to offer practical and accessible contraceptive instruction to ever expanding networks of interested medical practitioners.[155] The group ensured that when the medical profession finally accepted that 'competence in contraceptive technique [was] an important part of the duties of the general practitioner', it could extend and disseminate its scientifically and medically comprehensive syllabus to universities and hospitals.[156]

After Wright's book, the FPA continued to review and revise its medical training curriculum. It developed unique specifications for accrediting doctors, midwives and nurses, and demanded ever more clinical experience and theoretical and conceptual proficiency prior to issuing certificates of competence. *Contraceptive Technique* remained an important reference and an efficient guide for medical practitioners to navigate the contraceptive options available in the 1950s. Its preeminence as a medical reference diminished from the 1960s,[157] due primarily to the emergence of oral contraceptives.[158] Finally, the FPA were engaged with the cultivation of one final knowledge network: voluntary marital, family and sex education and guidance societies including the Family Relations Group, National Marriage and Guidance Council and Eugenics Society and the medical and user networks within their exclusive reach. This networking reached a peak mid-century as the Royal Commission on Population was being undertaken in order to understand population health and decline, family mores, norms and ideals, and sexual and family planning practices, education and aspirations. The FPA and its education and counselling-based contemporaries appreciated this would have significant ramifications for their work and goals and each made sure to understand, engage with and inform the commission's activities and findings as far as was possible.

The Impact of Sex and Fertility Education and the Royal Commission on Population

Understanding population flux in modern Britain became a national interest as the Second World War came to an end. It became apparent, based on marriage, birth and death rates over the last century, that if drastic steps were not taken to replenish its twice devastated stock, the threat of zero population growth loomed. The birth rate, which peaked at 35 per 1,000 between 1865 and 1875, had slumped to '14.4 in the years before the war of 1939–45'.[159] To understand the definitively downward trend, a Royal Commission on Population was instituted on 2 March 1944 to grasp the phenomenon through three strands of inquiry: statistical, economic, and biological and medical.[160] Each had a broad mandate to examine current population movements, understand their causes and 'consider their probable consequences'; to determine and recommend what actions, if any, might be warranted to serve future national population interests.[161] The Royal Commission proved as important a turning point for medically sound and regulated sex and contraceptive education and provision, as it was for the general acceptance of family planning. Several prominent education-focused voluntary societies effectively exploited the inquiry to demand a sex and contraceptive education standard be set for the good of the British public.

The Royal Commission's findings were published in five volumes, one expressly focused on 'family limitation and its influence on human fertility'.[162] Through a comprehensive 'Family Census' questionnaire sent to

'1,400,000 married women in Great Britain', or ten per cent of the total married female population, attitudes and ideals regarding family size and construction were collected and analysed.[163] Additionally, physicians and nurses assessed modern fertility and contraceptive practices, to discover their impact on population decline.[164] The commissioners ultimately determined that fertility had not noticeably declined since the nineteenth century[165]; but rather, broader family planning and spacing knowledge and options had halved mean family sizes.[166] This was also attributed to 'social developments over the past seventy or eighty years [which] … accentuate the relative economic and other handicaps of parenthood', but it was suggested that changes in social policy and education (see below) might arrest or reverse this trend. Recent leisure-time pursuits in 'popular literature, the press, cinema, commercial advertising, the educational system' and the economy, were said to have 'lowered the [desirability] of the family in modern life'.[167] More and more young married couples were keen to spend their time and money on amusements rather than immediately undertaking all the obligations associated with responsible parenthood. The report found that the principles of responsible family planning had been imbibed by the masses and advised a similar 'educational effort [was also] needed to spread throughout the community some understanding of the broad facts of the population trend and its consequence'. Family limitation was now so popular and widespread that 'a wide development of sex education' and its incorporation into the national school curriculum was needed to positively advertise the 'practical crafts of homemaking',[168] and ensure that 'preparation for married life' taught that the pleasures and responsibilities of marriage had national as well as individual consequences.[169] The commissioners recommended that providing a standard, nationalised contraceptive and sex education through the school system offered the best chance of arresting population decline, whilst still allowing parents autonomy to space and time their families. This finally echoed the long-held views of both the Eugenics Society and NBCA/FPA.

The commissioners were conscious that the primary cause of the population decline being investigated was married couples intentionally manipulating the size and time they began and concluded their families. To fully understand this development, the Biological and Medical Committee formulated a large-scale statistical study of the 'mechanics of family limitation' to discover the complexities of the relationship between 'family limitation and its influence on human fertility during the past fifty years'. The sub-committee was comprised of prominent medical figures and affiliates of pro-contraception welfare societies; Margaret Jackson represented the Family Planning Association and Carlos Blacker represented the Eugenics Society. The investigation, run by the Royal College of Obstetricians and Gynaecologists, was conducted by circulating a detailed questionnaire on family limitation practices to over 50 infirmaries and hospitals and having affiliated doctors complete the surveys to guarantee uniformity. In May 1946 the Royal College Council approved the questionnaire which posed integral questions regarding family limitation and fertility:

1. How extensively is birth control practiced?
2. In what proportions are the different methods of birth control practiced?
3. Are there important differences between different social groups in the extent of the practice of birth control, or in the choice of a method?
4. To what extent is birth control, as practiced, effective?
5. What is the extent of involuntary infertility?
6. Does the practice of birth control affect the power to reproduce?
7. How important is abortion as a method of birth prevention?
8. What is the proportion of 'unplanned' pregnancies?
9. What is the proportion of 'unwanted' children?
10. What are the chief reasons given for using birth control?[170]

Several of these questions were central to the FPA's profile-raising and scientific agendas. As is discussed in Chapter 4, the association, and indeed its research-focused predecessor the BCIC, had long worked to answer these questions by using clinic case cards to collect sociological data and appointing a statistician to analyse it in order to inform and improve practice and acceptability.[171] The Royal Commission essentially ignored this previous work; commissioners discounting the BCIC and NBCA/FPA findings as lacking 'trustworthy answers' because 'patients attending a clinic [did] not represent an unbiased sample'.[172] Although this was true, the association research contained useful data about contraceptive efficacy and use, and already solicited patient's reasons for family planning and method preference. Further, as many association doctors also worked externally in a private capacity, their experiences were absolutely not limited to the poorest and least socially mobile classes, and could have effectively informed the broader investigation. Nonetheless, the commissioners ignored this expertise, to solely employ evidence gathered from the questionnaire based on four evidentiary strata:

(a) The respondent's sociological background.
(b) An all-inclusive reproductive history.
(c) general use and methods of family planning and motivations for use.
(d) Specific methods used, failures and abortions if method failure occurred.

Between August 1946 to June 1947, 11,078 British women completed the questionnaire, and to ensure that the survey included a realistic and representative population, were sourced from hospitals in industrial, urban and rural locations. This guaranteed that proportionate populations were considered in line with their overall representation.[173]

Within the questionnaire lay answers to current questions that were of global concern. Why couples were inclined to practice contraception? Whether contraceptives caused infecundity? And was there a distinction between classes and their contraceptive desire and use? Resolving these problems could help to understand current local and national population drifts as well as the pervasive appeal of the small family. The reasons revealed in the Royal Commission study were identical to those discovered a decade earlier in a comparable, but much smaller survey undertaken by Australian physician and FPA international affiliate Victor Hugo Wallace. Both studies determined there was no single reason people embraced contraception, but rather a combination of factors contrived to make smaller families most desirable to modern couples.[174] Some patients cited economic reasons such as the high cost of comfortably feeding, clothing, educating and housing a family. Other common reasons were medical, including the risk of maternity and possibility of transmitting genetic defects. Some cited social reasons, like a desire to first establish marital satisfaction, decreased religious authority prohibiting the practice, perceptions a small family was more proper, and female employment opportunities. Fear, presented as health, financial or employment risks, was also common. Finally, many admitted parenthood was not an aspiration.[175] Whilst these broad trends were common the specific application was personal, and Wallace's responses were indicative of this fact:

- My mother had ten children. Luckily, five of them died... I don't remember a happy day in my childhood, only starvation and misery.
- Nice people have only two children, or three at the very most.
- We could afford to have six children, but are not going to... unfortunately, they are all mentally retarded.
- I have been advised by several physicians that it would be highly dangerous, owing to my serious heart condition, to become pregnant.
- We believe that a period of about two years is desirable [after marriage] before the birth of the first child.
- The doctor said he nearly lost me at the last one and advised me not to have any more children. [Also] my husband and I think that, for us, two children are sufficient to educate and bring up properly.
- I'm not the sort of girl who likes to stay at home... I like to go out and see a bit of life. I'm interested in my job. Personally, I'm not at all keen on having a baby.
- I am a State School Teacher... I am allowed to get married... but I must not have a baby. If I did I would forfeit the Education Department £200 [bond].[176]

The statistical data collated by the Royal Commission and Wallace's questionnaires agreed that the 'population problem' that was now afflicting both nations, was mostly immaterial to individual's reproductive lives and decision making. Such public sentiment had and would continue to play a 'not unimportant part' in population decline. It was recognised there existed major impediments to increasing national fertility that included increased divorce rates, diminished 'home industry', a mass-migration induced loss of community and social networks, economic hardships and the actual and psychological effects of successive world wars. Contraception had little to no influence on any of these issues. The authors of each study admitted that in the current cultural climate it would be futile to attempt to re-educate couples that their sex lives were vital to national interest. Indeed, most now subscribed to the ideal of voluntary parenthood and believed overpopulation loomed, and so respondents justified personal fertility choices as conscientious social decisions with positive lifestyle benefits. This psychosocial phenomenon wherein couples considered a small family and 'personal interests and comfort' to be essential life aims, was condemned as 'selfish and irresponsible' by many medical men and tacitly, the commissioners.[177] Ironically, eugenists and contraceptive advocates called this decision 'responsible parenthood' and it was the desired outcome of their separate and combined industry.[178]

Competing efforts to raise the birth rate and dissuade couples from unplanned parenthood peaked mid-century as the contraceptive movement flourished and the necessity of population growth became undeniable. The best means to effect both was through public education programs on reproductive science, marital relations and hygiene, sexual health and contraception. As was described, lessons addressing each of these had been pioneered by the NBCA and Eugenics Society, in partnership or individually, and in its submission to the Royal Commission, the FPA offered these syllabi to the Ministry of Health and Board of Education, who it suggested could widely and effectively standardise and propagate the knowledge.[179]

The research undertaken for this inquiry was based on two assumptions: that most people were aware of contraception, and that up to 35 per cent of couples married since 1940 actively employed contraceptive techniques and devices. This was borne out by statistics proving that between 1900 and 1947 there was a sharp rise in exposure to and use of contraception across all classes. This phenomenon has recently been examined by Claire Jones and Jessica Borge in relation to contraceptive manufacturing, advertising and promotional activities; they report a huge uptick among moderate and unskilled classes. Over half a century awareness rose from four to 47 per cent. These statistics demonstrate that the NBCA/FPA and Eugenics Society's decades long promotional and educational work, in concert with permissive cultural and commercial shifts, had proved effective. In fact by the mid-1930s all classes received approximately equivalent exposure to contraceptive knowledge and products for purchase.[180] The result was that the paired ideals of voluntary parenthood and family planning were actively applied in modern marriages.[181] Although other social, economic and personal factors contributed to downward population trends, it was apparent the public exhibited an expanding desire to understand and control fertility, and were well aware of where and how that goal could be attained: FPA clinics.

Unfortunately, for historians interested in public access to contraceptive knowledge, the Royal Commission questionnaire had one distinct failing: it paid no focus to how and where participants became acquainted with contraceptive knowledge. In *Sex Before the Sexual Revolution* (2008), Simon Szreter and Kate Fisher undertook oral histories to specifically interrogate the 'culture of sexual ignorance' defined and argued by earlier historians.[182] They found that the ways and means that people were exposed to sexual information were extremely varied and that the perceived sexual and contraceptive ignorance of mid-twentieth century women was not universal. I contend that significant scientifically and physiologically sound knowledge of sex and contraception was widely available by 1930. This is based on the responses of Royal Commission interviewees, well-documented early-century letters to Marie Stopes demanding the democratisation of sex education,[183] the general knowledge of contraception claimed by the BCIC in 1928 and then proved by the NBCA's applied scientific studies that showed the practice was 'widely used' but that medical and scientific investigation was needed to guarantee the methods and technologies, and finally, Jones and Borge's studies of the flourishing contraceptive manufacturing industries, print cultures and sales and consumption of products, books and indeed trade catalogues.[184] I assert that the association and Eugenics Society both used official aversion to and ignorance of contraceptive and sex education to contrive and prolong an aura of ignorance that heightened the urgency of their joint calls for a 'universal accessibility of [sexual and contraceptive] knowledge'.[185]

To claim that this existing knowledge was of a valuable caliber for comprehension or application would be to misrepresent the findings of Kate Fisher's previous work *Birth Control, Sex and Marriage in Britain, 1918–1960* (2006) which details society-wide cultures and structures that 'suppressed and hushed up' matters of sex. These included government and medical authorities' that refused sex and contraceptive instruction except for medical need, educational powers' that denied any responsibility, and finally, families and social groups that colluded to perpetuate gender-based sexual knowledge inequality. Complete illiteracy was the female ideal, and it was guaranteed through the use of euphemism when these topics emerged, whereas for men, frank, crude group discussions were a common source of sexual and contraceptive guidance.[186] This gendered sex education, which relied on male partners to gather and dispense such knowledge to their partners, shifted after 1920. The Royal

Commission enquiry discovered that between 13 and 27 per cent of women married before 1920 (depending notably on class) still declared themselves 'ignorant of [contraceptive] steps that could be taken'. But amongst women married in the interwar years, when the NBCA formed and forcefully promoted its services chiefly to the middle and lower classes, claims of ignorance dramatically reduced to seven per cent in professional and skilled classes and 16 per cent amongst the unskilled. A parallel decline was noted regarding religious objections to contraception, as well as perceptions that contraceptives were unsafe.[187] This clearly suggested a growing acceptance and trust for contraception amongst all social strata and bolsters my assertion that the NBCA/FPA's publicity and education efforts were broadly successful. The Royal Commission did not ask, but this public trust likely resulted from amplified association-linked exposure to contraceptive methods and advice (through sites, its affiliated manufacturers and medical and user word-of-mouth), as well as its medical clinical practices and official guarantees that only scientifically and medically tested and approved products were dispensed. As users reported diminished anxiety regarding contraceptive technologies, a contrary perception emerged regarding failure to plan families. This came in the wake of women's movement-led cultural shifts that asserted 'it would [actually] be harmful to attempt to restrict the contribution that women [could] make to the cultural and economic life of the nation', and that they should 'combine motherhood and the care of a home with outside interests'.[188] The Royal Commission ultimately revealed that the huge social, political and economic upheavals of the last century had affected a near-universal ideal that constant maternity was a physical and social burden to be avoided for health and happiness. This message that had long been the FPA's central argument to support its contraceptive efforts reverberated widely and became central to the educational lobbying efforts of other voluntary marriage and family focused societies in the decade after the commission released its findings.

Uniting Pro-family Planning Voluntary Efforts and Lobbying to National Educational Effect

The Royal Commission was a boon for publicising the association and its broad and successful contraceptive and medical services, scientific research, and educational work to the still largely ignorant adult public.[189] In finding that the newly founded National Health Service

(NHS) should provide 'advice on contraception to married persons who want it' and that all restrictions on giving sound contraceptive advice 'should be removed', the Royal Commission also bolstered the association's long-held goal that its work be transferred to the Ministry of Health. This led the FPA to seek Ministry funds 'in the region of £20,000 per year' to extend its clinical and instructive work.[190]

Buoyed by the Commission's support for family preparation education, the association and its progressive medical affiliates encouraged school-based sex and reproduction lessons for elder female students via 'actual instruction and demonstration of infant welfare'.[191] It also publicly argued for practical and theoretical preparation and instruction of school age children and young people about the grave responsibilities of parenthood. It's worth noting at this juncture that the FPA perceived a district difference between practical and age-appropriate instruction on sex and contraception. The FPA endorsed 'home-making courses, [and] adult-class extensions of marriage psychology', to complement the proposed 'wider development of sex education' for high-school girls.[192] But internal discussions and documents regarding practical instruction compiled by Wright and her son Beric, directed that 'the best time for introducing a girl to contraceptive technique is about three weeks before her marriage'.[193] Thus gradations of education and practical instruction were being devised and promoted that catered to the various and specific needs of audiences based on sex, age and life and family circumstances. The Board of Education disregarded the FPA's calls for it to take a more active role and announced its position on sex education had been communicated in *Sex Education: In Schools and Youth Organisations* (1943). The pamphlet claimed that school-based sex education was unwarranted as the necessary biological reproductive knowledge 'comes to everyone sooner or later', and that parents and guardians were best placed to teach such matters through candid dialogue. The pamphlet optimistically concluded that children taught this way 'will require no set instruction in the physiology of sex as a *particular* topic at a later stage'.[194] This conclusion was of course nonsensical when viewed within the pervading Foucauldian culture of sexual secrecy that paradoxically worked by simultaneously limiting and facilitating the production and dissemination of sexual knowledge. Understanding sex and its consequences was deemed vital knowledge for every adult, but only knowledgeable adults were allowed to impart it. There was clearly a step missing in the necessary knowledge network, and even the Education Board admitted: a 'substantial proportion' of parents did not possess the inclination, vocabulary or basic knowledge to explain sex and reproduction. Hence 'many schools' without a formal mandate 'felt it their duty, to undertake the very important task' of correcting 'the inadequate and distorted knowledge possessed by many children'.[195] The generational cohort that was receiving this newly articulated scientific sexual knowledge would likely be competent to instruct and guide its own future children, but it certainly wasn't a viable option mid-century. Still, the Board of Education kept compulsory sex education off the national syllabus, and instead proposed that local authorities might provide short courses for teachers and youth leaders and hold meetings to guide parents in such discussions.[196]

This tacit Board of Education support and the Royal Commission's assertion that 'nothing should … impede the spread of family limitation knowledge' fortified the FPA and its funder the Eugenics Society's resolve to achieve their shared goals of passing tested and standardised sex and contraceptive lessons on to some national authority.[197] However these were not the only local providers of this education, and indeed, as I have discussed above and Rusterholz and Jones have each pointed out, even in its most obvious field of expertise—contraception—the FPA was not the sole educational expert or source. Other groups provided educational programs spearheaded, designed and taught by their own sex and contraceptive experts: the Central Council for Health Education, National Marriage Guidance Council and Family Relations Group. It was in the wake of the Royal Commission that these groups began to actively align their disparate efforts to effect jointly beneficial educational aims. The conclusion of this chapter discusses the FPA's collaborative efforts to effect universal, scientific sex education.

At the 1946 inaugural meeting of the Family Relations Group it was agreed that the distinct British groups concerned with social welfare and family planning should cooperate regarding the five integral 'paramedical' matters under their collective focus: the treatment of infecundity, contraception, eugenic problems, pre-marital examinations and marriage guidance. Although no issue was then entirely ignored by the public health service, each local health authority's management was so disparate that these matters were 'more the concern of voluntary organisations than statutory bodies', and as such, there was now an opportunity for the latter to definitively impact the way these were understood and addressed. This new medico-social alliance would function like the early NBCA to develop a 'practical scheme or programme' to provide various family services understanding that the need for assistance for most matters under their concern was now, and would hopefully remain, small. The issues demanding expert education and advice would initially be marriage guidance and contraception, but it was expected that infertility treatment and guidance would be incorporated as fertility assistance was increasingly sought at FPA sites.[198] The new collective recommended the FPA launch an infecundity clinic to service these patients and complement its marital difficulties sessions and seminolgical testing work to understand and assess male sterility.[199] This became and remained a peripheral FPA focus.

The Family Relations Group was able to effect and suggest a broad research and service remit because its board included representatives from the FPA, British Social Hygiene Council, and Eugenics Society amongst other welfare collectives, and hosted observers representing the Home Office, Ministry of Education and Ministry of Health. It was a new education-focused association and in that capacity conceded that the FPA was already somewhat addressing each of its five stated interests, but it flagged that its intention was to use its broad network's diverse experience and expertise to expand and heavily promote each field of knowledge.[200] Through these five medico-scientific and educational foci the group and its allies could facilitate both medical and scientific research and knowledge collection and export, and also expand the sex education lobby's aptitude for publicising and providing the public knowledge regarding normal and abnormal reproduction and fertility, as well as family planning and limitation methods and practices.[201] All the educational and therapeutic programs founded by the FPA and adopted by the new group required a high standard of general sex literacy to foster self-awareness, as well as trust in the quality and methods of instruction on offer. Once it became apparent that the Family Relations Group would be the best placed organisation to achieve the association's educational goals, the FPA and its staff actively provided guidance and expertise to that end.

To advance its key goal of understanding the nation's family dynamics and desires, the group commissioned research into its affiliates 'para-medical' work. The FPA was asked to report on contraception and infecundity, the Eugenics Society on pre-marital health examinations and eugenic problems, and the FPA's rival in the provision of this service, the National Marriage Guidance Council to assess the current state of marriage guidance. This research agenda was directly inspired by the 1946 institution of the NHS. The Act had huge ramifications for families, including providing a family allowance and extending health amenities to mothers and children, and as a result all these societies anticipated the NHS would imminently subsume most of the services each offered.[202] FPA General Secretary Margaret Pyke clarified

> Under the N.H.S. Act Local Authorities are specifically empowered to support voluntary associations doing 'preventative work' ... [and so] sooner or later some Government will have to take, or can be forced into taking, ... action to provide advice on family planning for the married people of this country.[203]

After the Act was introduced, these societies worked to unify and standardise their educational and welfare facilities to streamline the anticipated handover. By 1950, the FPA annually reviewed and guaranteed the products and methods it prescribed through its approved list, employed a regulated clinical practice operating within accepted medical pedagogies, had advanced pregnancy and sterility testing and diagnostic facilities and ran a popular medical contraceptive training program for doctors and nurses. All that remained to be formalised and regulated was a comprehensive sex education agenda. The FPA admitted that its already stretched staff had 'not the time [them]selves to do all these jobs' and this was the reason it sought and cultivated educational partners.[204] Fortuitously, during the 1950s two prominent groups with a sex, marriage and contraceptive education focus assumed this work and caused it to finally fall within the Board of Education's remit: the National Marriage Guidance Council and the Central Council for Health Education.

The FPA Eugenics Society, Family Relations Group, National Marriage Guidance Council and Central Council for Health Education all guided the British public to make wise marriage, sex, family size, and life choices, and agreed they had a 'clear duty to the next generation to provide ... instruction and guidance [to] safeguard them from wrong attitudes and false judgements'.[205] The groups did not ever formally unite, but closely cooperated to achieve shared ends. Each assigned delegates to represent its specific interests to the other associations and collaborated to define united education principles and methodologies. But even when they formally co-operated as they did to petition the Royal Commission on Population and the 1951–1955 Royal Commission on Marriage and Divorce, each was careful not to impede any other's autonomy or expertise. This was odd given that the groups agreed that voluntary societies with shared aims, but uncoordinated means, provided 'the best excuse' for official inaction on important but yet avoided public health issues.[206]

In 1938 a medical and clerical collective founded the National Marriage Guidance Council to guide young people to achieve happy partnerships and families, prepare them for the realities of sex and safeguard the institution of marriage. The next year the group launched its first lecture courses in central London to assist those about to marry to manage expectations, realities and difficulties, and address the concerns of those already married—a near identical service to those then being founded by the FPA, except with a slightly greater focus on Christianity. The lectures were a roaring success, starting with 150–200 attendees and increasing markedly. This encouraged the council to expand its lecture series throughout Britain. In the 1940s it published guides for the various stages of marital life, 'recommended books on various subjects likely to arise', and warned of the dangers of venereal disease and prostitution. To reinforce its popular lecture series, it also encouraged the post-war nation to 're-establish family ties' by 'rebuilding relationships between women and children and returning male combatants' and introduced counselling services to assist couples experiencing this difficulty.[207] Given the extreme overlaps between these services and those offered by competing organisations, its members also convened a conference for physicians, officials and other marital guidance groups to formulate standard programs and goals.[208] This work, like much of that of the association, was critically assisted by the award of Eugenics Society funds, as the groups had long claimed that marriage guidance had 'definite eugenic importance' and that a 'eugenic attitude [toward] marriage and the family' should be central to such educational work.[209]

Cohesion between the FPA, Eugenics Society and National Marriage Guidance Council and their educational and promotional agendas and methods was promised when FPA President Lord Thomas Horder was appointed as joint President of the latter council in 1943. From that position he rather unsuccessfully tried to effect cooperation by encouraging unity, at least, on marital education. Each group resisted. The Eugenics Society, like the government, openly preferred the FPA's scientific marriage guidance to the Marriage Guidance Council's religious methodology, and further appreciated the association had a vast network of British clinics from which to preach.[210] However, the society perceived a 'wonderful opportunity of reaching [and tailoring the same information to diverse] public' audiences was promised by collaborating with each. But Horder's lengthy experience bidding the medical profession and government to assume contraceptive work, gave him pause about the usefulness of a formal alliance. He astutely recognised that the medical profession's enduring affiliation with the National Marriage Guidance Council, might assist it to sway the new Central Medical Board that administered general practitioners within the NHS, to similarly embrace the FPA and Eugenics Society-backed scientific marital and sexual education.[211] Medical support would legitimise marital guidance as a therapy and necessitate the incorporation of sex and contraceptive instruction into those programs. To this end, in 1947 Horder encouraged the council to amalgamate with the more comprehensive and aspirational Family Relations Group, to effect 'increasing standardisation and regimentation of many of [their separate] daily activities' to more clearly delineate and manage their joint and individual pursuits.[212] It was also hoped that incorporating the Ministry of Education-aligned Central Council for Health Education into this network might also encourage a formal acceptance of sexual and contraceptive health into the official youth and adult education systems. The voluntary network considered this to be an essential arena for future activism and planned to form a 'central council for family welfare' specifically to petition the Health and Education ministries, if the central council refused the offer.[213]

However, this concern was unwarranted as recent changes at the central council, specifically its 1942 amalgamation with the British Social Hygiene Council meant it had already assumed some duty for marriage guidance and sex education. This was initially effected as a Ministry of Health-directed wartime venereal disease awareness program.[214] The FPA perceived this amalgamation would expand its knowledge network and raise public sexual health and literacy standards, as collaborating with the renowned sex education-focused British Social Hygiene Council, would bring the material directly under the Ministry's purview. Further the appointment of Cyril Bibby, a former schoolmaster with broad sexual/social hygiene experience, as the council's education officer directly bolstered its focus on defining and deploying modern national sex education standards. Bibby forged ahead with this work in spite of the central council's initial hostility to his having a 'wider interest in sex education than just venereal disease'. Max Blythe's history of the council demonstrates that Bibby, against his employers will, actively exploited his position and affiliation to submit over two hundred sex education-based contributions to medical and education journals. He further rebranded and refocused the British Social Hygiene Council's journal as the *Health Education Journal* in order to more efficiently spread central council propaganda.

During his tenure at the central council, Bibby published his seminal instructive text *Sex Education: a Guide for Parents, Teachers and Youth Leaders* (1944). It aimed to remedy the national dearth of methodological comprehension and proficiency.[215] Through the council, Bibby initiated and directed Britain's first comprehensive and specifically age-appropriate sex education courses that covered the spectrum of sex, marriage and health issues, aimed to end stigma around their frank discussion, and guide future generations' approach to sex. To complement his guide, he charted the various ways people could be expected to receive sex education, from whom, and what attitudes ought to accompany such lessons.[216] This book defined the links between the FPA, Eugenics Society, British Social Hygiene Council and Central Council for Health Education's educational work, and promoted the necessary interrelation of science, medicine and public health to impart accurate sexual knowledge—at every life stage and for all marital, family and health situations. Responding to the 'radical and widespread change in public opinion' supporting sex education that followed the Royal Commission on Population,[217] the council undertook a sociological survey to gauge parent's attitudes to sex education in school; once it obtained a parental mandate the council's sex education programs began to be implemented. Much like the attitudes cultivated among members of contemporary voluntary knowledge networks, Bibby and the council's stance in relation to parents, schools and institutions was 'we are going to do this, and hope you will co-operate with us'. The resultant sex education program incorporated tailored 'specimen lectures' aimed at distinctive age and life stage groups: young people in co-educational format, older adolescent and young adult men, and separately women, and parents of secondary school children.[218] Practical activities were developed to teach youths biological and reproductive concepts; via animal breeding and care, physiological model building, and museum, 'botanical and zoological gardens, farms, fish hatcheries, [and] piggeries' visits. Programs for adults focused on teaching scientific, physiological, caring, medical and moral aspects of sex. The practical activity program included parentcraft, scientific experiments, microscopic observations and animal dissections, individual and group sex education tasks, discussions and debates, and viewings of informative films. These age-based schemes ran in conjunction with broad summer lecture courses for adults, teachers and youth leaders, as well as seminars deliberating moral and ethical aspects of such teaching.[219] The central council worked very closely with 'various organisations concerned with special fields of health education' like sex, marriage and contraception to build unified and standardised educational and training tools that wouldn't undercut or undermine others' sensitive and specific work, but rather to focus on solid fact and remove all question of morality from sex education.[220]

So, where did contraception fall on the councils' agenda? The matter was formally incorporated into Bibby's book but in a limited capacity. Family planning was important as it impacted sexual behaviour and socio-sexual mores, but like with eugenics, it had two opposing branches: positive and negative. The former allied closely with the FPA's lessons on family building and spacing, and the latter incorporated contraceptive practices that exclusively focused on preventing conception. As this contrast implied an inherent moral component to the practice, and the views of teachers, parents and/or students may conflict, strict conditions for providing only 'factual' advice were recommended. As the matter was already in the public domain in the form of widespread discussion and trade catalogues and adverts, the central council argued there was 'no valid reason why it should be withheld'. But, the extent of the knowledge the council advised betrayed its moral coquettishness on the subject: 'when you are… thinking of getting married yourselves, you'll be able to get more details from your doctor'.[221] The advice recommended for children, adolescents and young adults was essentially the same, that frankness was ideal, but discretion was always necessary in such lessons. The correct time and place for such instruction was for those about to marry and should come from a medical site—like a family planning clinic, specifically a medical clinician therein who was capable and experienced in education and prescription.

So, although at the close of the 1940s a body finally existed that boasted direct affiliation with the ministries of health and education, and actively attempted to define and implement a scientific and morally acceptable sex education for Britain, contraception remained peripheral. But with the central council's reach and renown came formal opposition from both medical teaching faculties that objected 'to the intrusion of non-medical people into their field', and educators 'who were astounded at the [Ministry of Education] … giving grants … in a course which included others than teachers'.[222] Despite this bluster, no formal opposition was recorded to sex education being located within the national scientific curriculum, as there was, after the Royal Commission, consensus that this topic was 'essential in ensuring the future health of the family, and therefore the nation'.[223] Notwithstanding this narrow political and educational achievement favouring certified sex instruction, the bureaucratic and administrative upheavals stimulated by the 1944 Education Act and the NHS Act (1948) caused the sex and contraceptive education lobbying network to lose momentum, and so, Bibby's *Sex Education* remained the 'accepted standard [instructive] text' into the 1970s.[224]

During this time the FPA continued to provide sex and contraceptive lectures, training and counselling in their 91 clinics around England, Scotland and Wales.[225] This early work is often downplayed by historians who stress the importance of the association's post-1968 educational expansion under new FPA Director Caspar Brook, despite his claim that he 'considered the educational aspect of family planning a marked failure'.[226] But I contend that as this chapter has demonstrated the association, its staff and ideological allies defined and circulated a high caliber of scientific sexual and reproductive education, that this condemnation must relate to the FPA's mid-century decision to defer some of its teaching to other bodies, rather than continue to direct already scarce resources and funds to services duplicated elsewhere.

The NBCA/FPA's relationship to sex and contraceptive knowledge networks was always formally respectful and, as far as was possible, collaborative. After the Royal Commission on Population, the FPA extended and widely advertised its sex and contraceptive instruction efforts to meet national need. From that time the association enthusiastically worked with and alongside other groups specifically the Eugenics Society, National Marriage Guidance Council, Family Relations Group, British Social Hygiene Council and Central Council for Health Education, to ensure and promote medicalised and scientific sex and contraceptive instruction that could be incorporated by the Health and Education Ministries and effectively taught to the masses.

Conclusion

This chapter has charted the association's educational interests and progress from its definition and standardisation of the basic sexual and physiological knowledge necessary for married people, to its collaborative educational lobbying and networking efforts in the aftermath of the Royal Commission on Population. I have shown that Wright and her female medical colleagues' practical educational work throughout the mid-twentieth century ultimately resulted in a heightened sex and contraceptive literacy for FPA and medical patients, and ultimately, the readers of her various manuals and guides. With little more than her own sexual and medical experiences that she was always willing to revise in the face of contradictory evidence, Wright wrote foundational sexological, educational and contraceptive texts for both lay and medical audiences. Her and the association's contribution to these fields cannot be overstated. Wright's political savvy and measured temperament allowed her to actively and effectively assist the association to slowly make inroads with the antagonistic medical profession, by developing, implementing and exporting a comprehensive medical training curriculum for practitioners. Following her lead, the association and its medical and executive staff worked tirelessly to exploit sociological, medical and demographic developments and investigations, and ultimately to forge and foster three distinct knowledge and education networks: user, medical and peer, to widely and effectively create, disseminate and promote age, class, professional and politically apt scientific sexual and contraceptive education for Britain.

Notes

1. Helena Wright, *Sex Fulfilment in Married Women: A Sequel to the Sex Factor in Marriage* (London: Williams and Norgate Ltd., 1948), 88.
2. Caroline Rusterholz, *Women's Medicine: Sex, Family Planning and British Female Doctors in Transnational Perspective, 1920–70* (Manchester: Manchester University Press, 2020); Hera Cook, *The Long Sexual Revolution: English Women, Sex and Contraception, 1800–1975* (New York: Oxford University Press, 2004); Barbara Evans, *Freedom to Choose: The Life and Work of Dr Helena Wright, Pioneer of Contraception* (London: The Bodley Head, 1984).
3. Obituary: Helena Wright', *British Medical Journal* 284, 6321 (1982): 1051; Evans, *Freedom to Choose*, 7.
4. Sophia Jex-Blake, *Medical Women: A Thesis and a History* (London: Oliphant, Anderson and Ferrier, 1886), 77–134; Evans, *Freedom to Choose*, 58–61.

5. George Ryley Scott, *Modern Methods of Birth Control, with a foreword by Sir Arbuthnot Lane* (London: Torchstream Books, 1933); Ettie A. Rout, *Safe Marriage: A Return to Sanity* (London: W.M. Heinmann [Medical Books] Ltd., 1922).
6. Evans, *Freedom to Choose*, 80.
7. Lesley A. Hall, 'Wright [nee Lowenfeld], Helena Rosa', *Oxford Dictionary of National Biography*, accessed 1 July 2022, http://www.oxforddnb.com/view/article/31859.
8. Evans, *Freedom to Choose*, 84.
9. Wright, *Sex Fulfilment*, 7.
10. Evans, *Freedom to Choose*, 112. Hall, 'Wright [nee Lowenfeld], Helena Rosa'.
11. Tina Johnson, *Childbirth in the Republican China: Delivering Modernity* (Plymouth: Lexington Books, 2011), 148–9. A vaginal fistula is a tear between the vaginal wall and bladder causing constant urine leakage.
12. Marie Stopes, *Mother England: A Contemporary History, Written by Those who have had No Historian* (London: John Bale, Sons and Daneilsson, LTD, 1929), 3, 18–9, 95, 111, 165–67.
13. Stopes' books were denounced for offering a 'practical guide to prostitution'. In 1922, Stopes brought a libel case against Halliday Sutherland for his claim that her contraceptives were harmful. The Catholic Church funded Sutherland's defence and the Anglican Church publicised his good character. Clive Wood and Beryl Suitters, *The Fight for Acceptance: A History of Contraception* (Aylesbury: Medical and Technical Publishing Co. Ltd., 1970), 165; Halliday Sutherland, *Birth Control: A Statement of Christian Doctrine against the Neo-Malthusians* (New York: PJ Kennedy and Sons, 1922); 'The Moral Problem of To-day', *Church Times*, 8 February 1924: 146.
14. Evans, *Freedom to Choose*, 128; Robert Jütte, *Contraception: A History*, trans. by Vicky Russel (Cambridge: Polity Press, 2008), 165–6.
15. Lara Marks, *Metropolitan Maternity: Maternal and Infant Welfare Services in Twentieth Century London* (The Netherlands: Rodophi B. V. Amsterdam, 1996), 138, 180–1.
16. Evans, *Freedom to Choose*, 129.
17. Audrey Leathard, *Fight for Family Planning: The Development of Family Planning Services in Britain, 1921–74* (London: The Macmillan Press, 1980), 29–30.
18. Evans, *Freedom to Choose*, 129.
19. Evans, *Freedom to Choose*, 131–2.
20. Evans, *Freedom to Choose*, 133.
21. NKWWC Medical Committee Minutes, 11 September 1936. Wellcome Library, Archives of the Family Planning Association (WL/SA/FPA), WL/SA/FPA/NK226.

22. Alison Bashford and Carolyn Strange, 'Public Pedagogy: Sex Education and Mass Communication in the Mid-Twentieth Century', *Journal of the History of Sexuality* 13, 1 (2004): 71–2.
23. Caroline Rusterholz, '"You Can't Dismiss that as Being Less Happy, You See it is Different." Sex Counselling in 1950s England', *Twentieth Century British History* 30, 3 (2019): 375–98; Lisa Featherstone, 'The Science of Pleasure: Medicine and Sex Therapy in Mid-Twentieth-Century Australia', *Social History of Medicine* 31, 3 (2019): 445–61; Hera Cook, 'Getting "Foolishly Hot and Bothered"? Parents and Teachers and Sex Education in the 1940s', *Sex Education* 12, 5 (2012): 555–67.
24. Sue Morgan, '"Iron Strength and Infinite Tenderness": Herbert Gray and the Making of Christian Masculinities at War and at Home, 1900–1940', in *Men, Masculinities and Religious Change in Twentieth-Century Britain*, eds. Lucy Delap and Sue Morgan (United Kingdom: Palgrave Macmillan, 2013), 183.
25. Personal Communication H. Wright, 28 July 1980. Quoted in Evans, *Freedom to Choose*, 140.
26. 'Anglican Lambeth Conference 1930—Marriage and Sex: Resolution 12', *Anglican Communion*, accessed 1 July 2022, http://www.anglicancommunion.org/resources/document-library/lambeth-conference/1930/resolution-12-the-life-and-witness-of-the-christian-community-marriage?author=Lambeth+Conference&year=1930.
27. 'Anglican Lambeth Conference 1930—Marriage and Sex: Resolution 15', *Anglican Communion*, accessed 1 July 2022, http://www.anglicancommunion.org/resources/document-library/lambeth-conference/1930/resolution-15-the-life-and-witness-of-the-christian-community-marriage?author=Lambeth+Conference&year=1930.
28. Evans, *Freedom to Choose*, 153.
29. Helena Wright, *Sex Factor in Marriage* (New York: The Vanguard Press, 1937), Preface.
30. Wright, *Sex Factor in Marriage*, 39; Evans, *Freedom to Choose*, 151.
31. Personal Communication H. Wright, 3 February 1981. Quoted in Evans, *Freedom to Choose*, 151–2.
32. Helena Wright, *What is Sex?* (London: Williams and Norgate Ltd, 1947), 9.
33. Wright, *Sex Factor in Marriage*, 39.
34. Wright, *What is Sex?*, 97.
35. Wright, *What is Sex?*, 28.
36. Wright, *What is Sex?*, 20–24, 52–58, 78–94.
37. Wright, *What is Sex?*, 98–9, 102, 104–5.
38. NBCA Medical Sub-Committee Minutes, 5 July 1936. WL/SA/FPA/A5/88.

39. Syllabus for Lectures for Girls [c.1945]. WL/SA/FPA/NK/42.
40. Wright, *Sex Factor in Marriage*, 51.
41. Helena Wright, *More About the Sex Factor in Marriage* (London: Williams and Norgate Ltd., 1954), 26.
42. Martin Bush, 'The Rise of the Sex Manual', *History Today* (February 1999), 37.
43. Wright, *Sex Factor in Marriage*, 51.
44. Wright, *Sex Factor in Marriage*, 82–3. My italics.
45. Hera Cook, 'Sex and Doctors: the Medicalization of Sexuality as a Two-way Process in Early to Mid-Twentieth-century Britain', in *Cultural Approaches to the History of Medicine: Mediating Medicine in Early Modern and Modern Europe*, eds. Wiliem de Blecourt and Cornelie Usborne (London, Palgrave Macmillan, 2004), 193–4.
46. Wright, *Sex Fulfilment*, 23.
47. Wright, *Sex Factor in Marriage*, 11.
48. Wright, *Sex Fulfilment*, 17–8.
49. Wright, *Sex Fulfilment*, 44.
50. Wright, *Sex Factor in Marriage*; Wright, *What is Sex?*; Wright, *More About the Sex Factor*.
51. Wright, *More About the Sex Factor*, 48–9; Rusterholz, *Women's Medicine*, 115.
52. Wright, *Sex Fulfilment*, 74–6.
53. Wright, *More About the Sex Factor*, 89.
54. Diana Wyndham, *Norman Haire and the Study of Sex* (Sydney: University of Sydney Press, 2012), 420.
55. Caroline Rusterholz, 'Testing the Gräfenberg Ring in Interwar Britain: Norman Haire, Helena Wright, and the Debate over Statistical Evidence, Side Effects, and Intra-uterine Contraception', *Journal of the History of Medicine and Allied Sciences* 72, 4 (2017): 458–62.
56. Lesley A Hall, 'An Ignorance in Knowledge: Reflections on the History of Sex Education in Britain', in *Shaping Sexual Knowledge: A Cultural History of Sex Education in Twentieth Century Europe*, eds. Lutz D. H. Sauerteig and Roger Dawson (New York: Routledge, 2009), 20; Cook, *The Long Sexual Revolution*, 209–10.
57. Wright, *Sex Fulfilment*, 42.
58. The topic is mentioned once in relation to menopause in a revised edition of *The Sex Factor in Marriage*.
59. *The Social Workers Role in Family Planning*, 1970. Wellcome Library, Archives of the Abortion Law Reform Association (WL/SA/ALR), WL/SA/ALR/U48.
60. Norman E. Himes, 'Some Untouched Birth Control Problems', *Eugenics* 3, 2 (1930): 1–2.

61. NBCA Quarterly Letter to Members and Branch Secretaries, March 1937. Wellcome Library, Archives of the Eugenics Society (WL/SA/EUG), WL/SA/EUG/D/12/16/22.
62. NBCA Quarterly Letter to Members and Branch Secretaries, March 1937. WL/SA/EUG/D/12/16/22; Ministry of Health, *Memorandum 153/M.C.W* (London: Ministry of Health, July 1930); Ministry of Health, *Circular 1208* (London: Ministry of Health, 1931); *NBCA Sixth Annual Report, 1935–1936*, 1937. WL/SA/EUG/D/12/16/19; 'Incidence of Abortion' NBCA Report for the Ministry of Health, 1937. WL/SA/EUG/D/12/16/19; 'Fewer Britons', *The Times (London)*, 2 March 1939; 'Social Aspects of Birth Control', *The Times (London)*, 30 November 1932; *NBCA Subscription or Donation Flyer* [c.1937]. WL/SA/EUG/D/12/16/19.
63. 'Literary Supplement', *The Times* (London), 2 March 1939.
64. Functions of the Medical Sub-Committee, 1934. WL/SA/FPA/A5/88.
65. NBCA Medical Sub-Committee Minutes, 3 March 1935. WL/SA/FPA/A5/88; FPA Medical Sub-Committee Minutes, 23 November 1941. WL/SA/FPA/A5/88.
66. NBCA Medical Sub-Committee Minutes, 21 July 1935. WL/SA/FPA/A5/88; NBCA Medical Sub-Committee Minutes, 6 October 1935. WL/SA/FPA/A5/88.
67. NBCA Medical Sub-Committee Minutes, 31 January 1938. WL/SA/FPA/A5/88.
68. Carlos Blacker, *Birth Control and the State: a Plea and a Forecast* (London: K. Paul, Trench, Trubner, 1926); Carlos Blacker, *The Chances of Morbid Inheritance* (London: Lewis, 1934); Gladys M. Cox, *Clinical Contraception*, Second Edition (London: William Heinemann [Medical Books] Limited, 1937); Norman Himes, *Practical Birth-Control Methods, Revised by Margaret Jackson* (London: Allen and Unwin, 1940); Joan Malleson, *Any Wife or any Husband: a Book for Couples Who Have Met Marital Difficulties and for Doctors* (Harmondsworth: Penguin, 1962); Helena Wright, *Birth Control: Advice on Family Spacing and Healthy Sex Life* (London: Cassell's Health Handbooks, 1935); Helena Wright, *Contraceptive Technique: A Handbook for Medical Practitioners and Senior Students* (London: J&A Churchill Ltd., 1951).
69. Recommendations as to the Future Relation of the NBCA to the Eugenics Society, April 1937. WL/SA/EUG/D/12/16/19.
70. Wright, *Birth Control*, 5, 12–3.
71. NBCA Medical Sub-Committee Minutes, 16 October 1935. WL/SA/FPA/A5/88; Wright, *Birth Control*, 7; Lucy Bland and Leslie A. Hall, 'Eugenics in Britain: The View from the Metropole', in *The Oxford Handbook of the History of Eugenics*, eds. Alison Bashford

and Phillippa Levine (Oxford: Oxford University Press, 2010), 218–9; David Doughan and Peter Gordon, *Dictionary of British Women's Organisations, 1825–1960* (United Kingdom: Routledge, 2013), 48–9.

72. Wright, *Birth Control*, 7–9, 17.
73. Wright, *Birth Control*, 23–4.
74. Wright, *Birth Control*, 12, 27.
75. Wright, *Birth Control*, 28–9, 31–2.
76. Wright, *Birth Control*, 11–2, 24–5.
77. Wright, *Birth Control*, 52.
78. Wright, *Birth Control*, 51.
79. Claire L. Jones, *The Business of Birth Control* (Manchester: Manchester University Press, 2020), 152–3.
80. Bashford and Strange,'Public Pedagogy', 84, 90.
81. Wright, *Birth Control*, 51–5.
82. Wright, *Birth Control*, 68.
83. Other contemporary books on contraception included Joan Malleson's *Principles of Contraception*, Michael Fielding's *Parenthood: Design or Accident*, Marie Stopes', *Married Love*, *Planned Parenthood* by Mary Denham, and the pamphlet *Marital Hygiene*, *Modern Marriage and Birth Control* by Edward Griffith.
84. Wright, *Birth Control*, 75–6.
85. Wright, *Birth Control*, 83.
86. Wright, *Birth Control*, 88.
87. General Practitioner's Knowledge of Contraceptive Technique [c.1937]. WL/SA/FPA/A19/1.
88. Leathard, *Fight for Family Planning*, 16.
89. Rusterholz, *Women's Medicine*, 17, 40, 45–6.
90. Annotated List of Hospitals with Medical Schools, 1934. WL/SA/FPA/A19/1.
91. James Merchant, ed., *Medical Views on Birth Control, with an Introduction by Sir Thomas Horder* (London: Martin Hopkinson and Co. Ltd., 1926).
92. Charles Gibbs and Arthur Newsholme, *Medical Aspects of Contraception: Being the Report of the Medical Committee Appointed by the National Council of Public Morals in Connection with the Investigations of the National Birth-rate Commission* (London: Martin Hopkinson and Co., 1927), 78–9.
93. Gibbs and Newsholme, *Medical Aspects of Contraception*, 81.
94. Norman Haire, 'A Critical Review of Medical Aspects of Contraception', *New England Journal of Medicine* 200, 1 (1929): 13; 'Medical Aspects of Contraception', *British Medical Journal* 2, 3489 (1927): 952; Carol R. McCann, *Birth Control Politics in the United States, 1916–1945* (Ithaca: Cornell University Press, 1994), 85.

95. Gibbs and Newsholme, *Medical Aspects of Contraception*, 35–6.
96. Gibbs and Newsholme, *Medical Aspects of Contraception*, 12.
97. 'Medical Aspects of Contraception', *British Medical Journal*, 953.
98. 'Medical Aspects of Contraception', *Public Health* 41 (1927–1928): 101; 'Medical Aspects of Contraception', *Lancet*, 2, 5442 (1927): 1325–6.
99. International Medical Group for the Investigation of Birth Control Report, 1928. WL/SA/EUG/D/12/15.
100. Roger N. Goodman, 'Clinics for Birth Control', *Lancet* 1, 5816 (1935): 400.
101. NBCA Executive Committee Circular Recruiting Membership for Training Committee [c.1932]. WL/SA/FPA/A19/1.
102. Medical Training [c.1932]. WL/SA/FPA/A19/1.
103. Medical Training [c.1932]. WL/SA/FPA/A19/1.
104. Evans, *Freedom to Choose*, 144.
105. June Rose, *Marie Stopes and the Sexual Revolution* (London: Faber and Faber, 1992), 144–5; Medical Training [c.1932]. WL/SA/FPA/A19/1.
106. Rose, *Marie Stopes and the Sexual Revolution*, 204.
107. Sub-Committee on Training Report, 28 February 1933. WL/SA/FPA/A19/1.
108. Letter M. Pyke to E. Fuller, 24 February 1933. WL/SA/FPA/A19/1.
109. Response E. Fuller to M. Pyke, 26 February 1933. WL/SA/FPA/A19/1.
110. Letter M. Pyke to E. Fuller, 30 March 1933. WL/SA/FPA/A19/1.
111. NBCA and SPBCC Meeting Minutes, 23 June 1932. WL/SA/FPA/A19/1.
112. Comments on NBCA Training Scheme [c.1933]. WL/SA/FPA/A19/1.
113. Medical Training [c.1932]. WL/SA/FPA/A19/1.
114. List of Gynaecologists Written to by Lord Horder [c.1933]. WL/SA/FPA/A19/1; List of Hospitals with Medical Schools, 1934. WL/SA/FPA/A19/1; Letter NBCA Headquarters' Organiser to M. Rice, 21 July 1933. WL/SA/FPA/A19/1.
115. Letter NBCA Headquarters' Organiser to Lord Horder, 23 June 1933. WL/SA/FPA/A19/1.
116. Form Letter Lord Horder to Post Graduate Colleges, 11 May 1934. WL/SA/FPA/A19/1.
117. Letter NBCA Headquarters' Organiser to M. Rice, 21 July 1933. WL/SA/FPA/A19/1; Personnel of the Medical Curriculum Committees [c.1933]. WL/SA/FPA/A19/1.
118. Letter Fellowship of Medicine and Postgraduate Medical Association to Lord Horder, 28 October 1933. WL/SA/FPA/A19/1.
119. Letter T. I. Bennett to Lord Horder, 7 September 1933. WL/SA/FPA/A19/1.

120. M. Pyke, 'Contraceptive Methods', *Lancet* 2, 5857 (1935): 1075; M. Pyke, 'Contraceptive Technique', *British Medical Journal* 2, 3908 (1935): 1075; General Practitioner's Knowledge of Contraceptive Technique [c.1937]. WL/SA/FPA/A19/1.
121. NBCA Medical Sub-Committee Minutes, 26 January 1936. WL/SA/EUG/D/12/16/22.
122. NBCA Medical Sub-Committee Minutes, 21 July 1935. WL/SA/EUG/D/12/16/22.
123. NBCA Sub-Medical Committee Minutes, 6 October 1935. WL/SA/EUG/D/12/16/22.
124. Telephone Note P. Cripps to R. Peers, 2 February 1952. WL/SA/FPA/A19/2.
125. Letter NBCA to University of Cambridge Secretary of Examiners, 4 January 1935. WL/SA/FPA/A19/1.
126. Leathard, *Fight for Family Planning*, 57.
127. NBCA Executive Committee Minutes, 8 May 1936. WL/SA/FPA/A5/2.
128. NKWWC Doctors Committee Notes, 6 July 1934. WL/SA/FPA/NK224.
129. Advertisement 'A Lecture on the Theory and Practice of Contraception', 3 March 1933. WL/SA/FPA/A19/1.
130. FPA Standards for Training Clinics, December 1950. WL/SA/FPA/A19/1; Medical Training [c.1932]. WL/SA/FPA/A19/1; NKWWC Seventeenth Doctors Committee Report, 2 August 1934. WL/SA/FPA/NK225; NBCA Medical Sub-Committee Minutes, 31 January 1938. WL/SA/EUG/D/12/16/22; NKWWC Doctors Committee Report, 2 August 1934. WL/SA/FPA/NK224.
131. Letter L. Arnold to H. Wright, 28 September 1949. WL/SA/FPA/A19/1.
132. NKWWC Medical Committee Minutes, 6 September 1949. WL/SA/FPA/NK227.
133. Letter M. Pyke to M. Nicholson, 12 July 1949. WL/SA/FPA/A19/1.
134. Letter M. Pyke to M. Nicholson, 15 July 1949. WL/SA/FPA/A19/1.
135. NBCA Ad-Hoc Training Clinic Committee Minutes, 8 March 1951. WL/SA/FPA/A19/2.
136. Points to be Attached to Memorandum on Training Clinics, 6 March 1952. WL/SA/FPA/A19/2.
137. Draft Memo to Training Clinics, November 1949. WL/SA/FPA/A19/1.
138. Letter M. Pyke to M. Rice, 27 October 1949. WL/SA/FPA/A19/1.
139. NBCA Medical Sub-Committee Resolutions, 24 January 1950. WL/SA/FPA/A19/1.

140. Letter P. Crips to A. Court, 3 October 1952. WL/SA/FPA/A19/2.
141. Letter P. Crips to M. Salmon, 15 October 1952. WL/SA/FPA/A19/2.
142. Proposed Draft of FPA Training Standards, December 1950. WL/SA/FPA/A19/1.
143. Royal Commission on Population, *Report of the Royal Commission on Population* (London: Her Majesty's Stationery Office, June 1949), 194.
144. Wright, *Contraceptive Technique*, v.
145. Wright, *Contraceptive Technique*, 1.
146. Wright, *Contraceptive Technique*, 7–8.
147. Wright, *Contraceptive Technique*, 10.
148. Jones, *The Business of Birth Control*, 207; Wright, *Contraceptive Technique*, 20.
149. Wright, *Contraceptive Technique*, 48–55.
150. Royal Commission on Population, *Report of the Royal Commission on Population*, 5.
151. Wright, *Contraceptive Technique*, 64–5.
152. NKWWC Medical Committee Minutes, 26 November 1951. WL/SA/FPA/NK227.
153. Beryl Suitters, *Be Brave and Angry: Chronicles of the International Planned Parenthood Federation* (London: International Planned Parenthood Federation, 1973), 60–64, 143.
154. Evans, *Freedom to Choose*, 225–9.
155. Gibbs and Newsholme, *Medical Aspects of Contraception*, 12, 36.
156. Wright, *Contraceptive Technique*, 63.
157. Evans, *Freedom to Choose*, 232.
158. Training Courses in Contraceptive Techniques for Nurses, [c.1966]. WL/SA/FPA/A19/14.
159. 'Report of the Royal Commission on Population', 76, 38.
160. I. B. T., 'Great Britain Royal Commission on Population Papers', *Population Index* 17, 2 (1951): 91–2.
161. Royal Commission on Population, *Report of the Royal Commission on Population*, 5; 'Report of the Royal Commission on Population', *Journal of the Institute of Actuaries* 76, 1 (1950): 38.
162. E. Lewis-Faning, *Papers of the Royal Commission on Population, Volume 1: Report on an Enquiry into Family Limitation and its Influence on Human Fertility During the Past Fifty Years* (London: His Majesty's Stationery Office, 1949).
163. Royal Commission on Population, *Report of the Royal Commission on Population*, 2.
164. Lewis-Faning, *Enquiry into Family Limitation*, 1–2.
165. Royal Commission on Population, *Report of the Royal Commission on Population*, 31.
166. Lewis-Faning, *Enquiry into Family Limitation*, 6–8.

167. Royal Commission on Population, *Report of the Royal Commission on Population*, 210.
168. Royal Commission on Population, *Report of the Royal Commission on Population*, 231.
169. 'Royal Commission on Population Report: Its Significance for the Future, by A Doctor', *New Health* (1949). WL/SA/FPA/A18/2.
170. Lewis-Faning, *Enquiry into Family Limitation*, 1–2.
171. BCIC Memorandum on Proposed Re-Organisation, 1930. WL/SA/EUG/D/12/12.
172. Lewis-Faning, *Enquiry into Family Limitation*, 1.
173. Lewis-Faning, *Enquiry into Family Limitation*, 3–5.
174. V. H. Wallace, 'Women and Children First: An Outline for a Population Policy for Australia', 1947. University of Melbourne Archives, Wallace Papers, (UOMA/WP), U41/15–19, Box 1, Files 1–3.
175. Wallace, 'Women and Children First', 43–63; Lewis-Faning, *Enquiry into Family Limitation*, 210–12.
176. 'Patient 6 Response', Wallace, 'Women and Children First', 44; 'Patient 40 Response', Wallace, 'Women and Children First', 58–9; 'Patient 27 Response', Wallace, 'Women and Children First', 54; 'Patient 2 Response', Wallace, 'Women and Children First', 45; 'Patient 4 Response', Wallace, 'Women and Children First', 44; 'Patient 36 Response', Wallace, 'Women and Children First', 57; 'Patient 20 Response', Wallace, 'Women and Children First', 51; 'Patient 14 Response', Wallace, 'Women and Children First', 47–8.
177. 'The Royal Commission on the Birth-Rate', *Sydney Morning Herald*, 5 March 1905; Stefania Siedlecky and Diana Wyndham, *Populate and Perish: Australian Women's Fight for Birth Control* (Sydney: Allen and Unwin, 1990), 17–8; Lewis-Faning, *Enquiry into Family Limitation*, 210–11.
178. Susan Burke, *Responsible Parenthood and Sex Education: Proceedings of a Working Group held in Tunisia, November 1969* (London: International Planned Parenthood Federation, 1969).
179. FPA Submission and Recommendations to Royal Commission on Population [c.1944]. WL/SA/FPA/A18/1.
180. Jones, *The Business of Birth Control*, 107–11.
181. Lewis-Faning, *Enquiry into Family Limitation*, 10–1.
182. Simon Szreter and Kate Fisher, *Sex Before the Sexual Revolution: Intimate Life in England 1918–1963* (Cambridge: Cambridge University Press, 2010).
183. Marie Stopes, *Dear Dr Stopes: Sex in the 1920s*, ed. Ruth Hall (London: André Deutsch, 1978).
184. BCIC Report of Sub-Committee on Re-Organisation. April 1928. WL/SA/EUG/D/12/12; Jones, *The Business of Birth Control*, chapters 2 and 5; Jessica Borge, *Protective Practices: The London Rubber*

Company and the Condom Business (Montreal and Kingston: McGill-Queen's University Pess, 2020), 135–49.
185. Eugenics Society Memorandum for Royal Commission on Population, January 1945. WL/SA/EUG/D/185.
186. Kate Fisher, *Birth Control, Sex and Marriage in Britain, 1918–1960* (Oxford: Oxford University Press, 2006), 44–62.
187. Lewis-Faning, *Enquiry into Family Limitation*, 176–7.
188. 'Royal Commission on Population Report: Its Significance for the Future, by A Doctor', *New Health* (1949). WL/SA/FPA/A18/2; Royal Commission on Population, *Report of the Royal Commission on Population*, 227.
189. 'The Family', *Manchester Guardian*, 21 June 1949. WL/SA/FPA/A18/2.
190. 'Family Planning', *Sussex Daily News*, 23 June 1949. WL/SA/FPA/A18/2.
191. Joan Malleson Submission to FPA Population Committee, June 1944. WL/SA/FPA/A18/1.
192. 'Future of Britain', *News Chronicle*, 21 June 1949. WL/SA/FPA/A18/2.
193. Draft for Discussion: A Handbook of Contraceptive Technique [c.1952]. WL/SA/FPA/A16/25/15.
194. Board of Education, *Educational Pamphlet No. 119, Sex Education in Schools and Youth Organisations* (London: His Majesty's Stationery Office, 1943), 1.
195. Board of Education, *Educational Pamphlet No. 119*, 22.
196. Board of Education, *Educational Pamphlet No. 119*, 1–2.
197. 'Royal Commission on Population Report: Its Significance for the Future, by A Doctor', *New Health* (1949). WL/SA/FPA/A18/2.
198. Family Relations Group Draft Memo, 4 January 1946. WL/SA/EUG/D/66. See also, WL/SA/FPA/NK/96 for patient sub-fertility correspondence between 1949 and 1951.
199. Letter Baroness Denman to A. Bevan, 22 July 1949. WL/SA/EUG/D/12/16/24.
200. Family Relations Group Conference on Education for Family Life, 5–6 January 1948. WL/SA/EUG/D/66.
201. The FPA—Past and Future: Speeches at the Annual General Meeting, 20 May 1950. WL/SA/EUG/D/12/16/24.
202. Cliff Alcock, Guy Daly, and Edwin Griggs, *Introducing Social Policy* (New York: Taylor and Francis, 2008), 391; Peter Alcock, *Social Policy in Britain* (Basingstoke: Macmillan, 1996), 281; The FPA - Past and Future: Speeches at the Annual General Meeting, 20 May 1950. WL/SA/EUG/D/12/16/24.

203. The FPA—Past and Future: Speeches at the Annual General Meeting, 20 May 1950. WL/SA/EUG/D/12/16/24.
204. The FPA—Past and Future: Speeches at the Annual General Meeting, 20 May 1950. WL/SA/EUG/D/12/16/24.
205. Memorandum on the Work of the National Marriage Guidance Council, 1939. WL/SA/EUG/D/120.
206. Letter A. J. Brayshaw to C. P. Blacker, 10 August 1951. WL/SA/EUG/D/184.
207. Jane Lewis, 'Public Institution and Private Relationship: Marriage and Marriage Guidance, 1928–1968', *Twentieth Century British History* 1, 3 (1990): 244–5.
208. National Marriage Guidance Council Report on Work, May 1938-February 1939. WL/SA/EUG/D/120.
209. National Marriage Guidance Council Report on Work, May 1938-February 1939. WL/SA/EUG/D/120.
210. Annotated Correspondence C. P. Blacker to Lord Horder, 8 July 1943. WL/SA/EUG/C/173.
211. Annotated Correspondence C. P. Blacker to Lord Horder, 31 March 1944. WL/SA/EUG/C/173.
212. Draft Family Relations Group Memorandum on Education for Family Life, 1946. WL/SA/EUG/D/66.
213. Annotated Correspondence C. P. Blacker to Lord Horder, 31 March 1944. WL/SA/EUG/C/173.
214. *Annual Report of the Central Council for Health Education 1944–5* (London: 1945), 19.
215. Max Blythe, 'A History of the Central Council for Health Education, 1927–1968' (PhD, University of Oxford, 1988), 186–8.
216. Cyril Bibby, *Sex Education: A Guide for Parents, Teachers and Youth Leaders*, Second Edition (London: Macmillan and Co. Ltd., 1954), 215–23.
217. Bibby, *Sex Education,* 1; Royal Commission on Population, *Report of the Royal Commission on Population*, 211.
218. Bibby, *Sex Education*, 266.
219. 'Medical News', *British Medical Journal* 1, 4343 (1944): 480.
220. 'Summer School in Health Education', *Health Education Journal* 2, 3 (1944): 100.
221. Bibby, *Sex Education*, 177.
222. Blythe, 'History of the Central Council', 264.
223. Padmini Iyer and Peter Aggleton, 'Seventy Years of Sex Education in Health Education Journal: A Critical Review', *Health Education Journal* 74, 1 (2015): 9.
224. Lesley A. Hall, 'Birds, Bees and the General Embarrassment: Sex Education in Britain, from Social Purity to Section 28', in *Public or Private*

Education? Lessons From History, ed. Richard Aldrich (London: Woburn Press, 2004), 106.

225. The FPA—Past and Future: Speeches at the Annual General Meeting, 20 May 1950. WL/SA/EUG/D/12/16/24; C. P. Blacker, *FPA Analysis of 43 filled-in Questionnaires. A Pre-war and Post-war Year Compared*, 1950. WL/SA/EUG/D/12/16/24.
226. Leathard, *Fight for Family Planning*, 151.

CHAPTER 4

Employing Pure and Applied Science to Assess Contraceptive Technologies

Abstract This chapter focuses on the pure and applied research projects that were instigated and directed by the National Birth Control Association/Family Planning Association and the North Kensington Women's Welfare Clinic. It contrasts the lab-based quality, efficacy and safety research undertaken at the association's direction, to argue that the development of the concept of contraceptive effectiveness and means of testing it, was a scientifically and politically fraught process that had to be carefully navigated. It also explores the association's efforts to implement conclusive testing methods to regulate and scientifically standardise chemical and mechanical contraceptives through the approved list. Further, it looks at the development of applied scientific projects which included the development of case cards and the exploitation of the clinic for undertaking pilot and later formal clinical trials to assess safety and efficacy of contraceptives prescribed in the clinic.

Allying contraceptive technology with the medical and scientific disciplines was not a straightforward task for the National Birth Control Association (NBCA)/Family Planning Association (FPA), despite its firmly held convictions and scientific research and education agendas and efforts. From its inception the association consistently undertook to make its sessions, lessons, methods and facilities more medical in practice, and public perception. In conjunction with the Birth Control

N. Szuhan, *The Family Planning Association and Contraceptive Science and Technology in Mid-Twentieth-Century Britain*, Medicine and Biomedical Sciences in Modern History, https://doi.org/10.1007/978-3-030-81300-0_4

Investigation Committee (BCIC) and the North Kensington Women's Welfare Centre (NKWWC), the association conceived and implemented a series of medico-scientific investigations to guarantee the contraceptive technologies and methods prescribed in its clinics.

This chapter charts the pure and applied scientific activities and tools that NBCA/FPA implemented to test, trial and certify its contraceptive products and services. It argues that through a series of sustained scientific laboratory- and clinic-based research, the association was able to guarantee the effectiveness and safety of contraceptive practices, methods and technologies. These ranged from comparative chemical investigations into the efficacy, safety and quality of spermicides, the publication of an approved list of contraceptives, and analyses of patient and clinic records to assess their acceptability, efficiency and safety in use. It further charts the origin and expansion of harmlessness testing from a peripheral regulatory complement to the association's chemical contraceptive testing, into a medical research specialty of contraceptive science. It also details the establishment and refinement of rubber contraceptive standards into the NBCA/FPA's regulatory repertoire and approved list.

The Approved List: Testing Contraceptive Effectiveness, 1928–1937, and Beyond

In 1933, within three years of its foundation, the NBCA established the 'Approved List' for chemical contraceptives.[1] This agenda and program remained prime to the FPA when it succeeded in 1939.[2] The premise of the approved list was that the NBCA would establish and ensure a base standard that every contraceptive product distributed by the association and used in its clinics, must meet.[3] The earliest existing copy of the list is dated October 1937. Under the auspices of the NBCA, a two-page list of products that 'have proved satisfactory in clinical practice and harmless in laboratory examination' was compiled. This list, plainly not yet intended for public consumption, continued 'they [the contraceptives] can be obtained from any good chemist' and asserted 'these manufacturers give special prices to clinics and doctors'.[4] This list was clearly projected to encourage the public to conflate the association and its function, products and services with the trusted and reputed fields of medicine and science.

Just fourteen products, manufactured by eight companies, were recognised on the list. These represented those successful contraceptive technologies for which modern scientific methods for testing efficacy and standardised manufacture had been developed by the 1930s. A selection of caps and sheaths were registered under the headline 'Rubber Appliances'. These represented the spectrum of female and male barrier methods which had emerged since vulcanisation had enabled rubber to be fashioned into durable 'thin and delicate sheets',[5] allowing the mass production of resilient water- and heat-proof caps and condoms.[6] The list incorporated 'womb veils' and 'female protectors' such as the vaginal sponge dipped in vinegar solution,[7] the diaphragm, initially dubbed the 'Mensinga Cap' or 'Dutch Pessary', and cervical caps, typically flexible, rubber cervical shields, produced quickly and cheaply in a variety of sizes. All of these methods ensured no sperm could enter the womb by blocking its entry. The diaphragm did so by covering the length of the cervix to the symphysis pubis, the cervical cap literally capped the cervix, and the sponge created a barrier and was doused in liquids that boasted spermicidal qualities. Condoms and sheaths were endorsed most enthusiastically as the 'check that is CERTAIN'.[8] Both worked by covering the erect penis to prevent sperm from entering the woman. The second and third sections of the approved list 'Pastes, Jellies, Etc' and 'Suppositories', featured products designed to be used in conjunction with these barrier methods. Soluble or melting pessaries, a 'small cone of cocoa butter charged with quinine' which melted at body temperature, or spermicidal paste or jelly, which was applied to a cap/diaphragm or inserted directly into the vagina shortly preceding intercourse to 'destroy the vitality of the seminal fluid'.[9] The association assured its consumers that 'all the goods mentioned [on the list] have passed the special tests of the NBCA'. It further promised that 'as clinical evidence accumulates this list will be emended [sic] and/or supplemented'.[10] In a 1950 expansion of the approved list, the association announced three tiers for contraceptive endorsement. Section A in the list featured 'pastes jellies and suppositories' that boasted a 'satisfactory level of spermicidal efficiency' and 'had been found reliable and harmless under prolonged clinical trial'.[11] Section B included products that had only passed laboratory tests but had yet to be proven 'under prolonged clinical trial'. Section C comprised products that had not passed lab testing, but had 'proved satisfactory and harmless' in use.[12] This publication became a defining feature and method of the association's drive to monopolise the British contraceptive market.

The NBCA/FPA, as industrial and engineering firms had done earlier, introduced a laboratory-supported set of testing and assessment standards for the technologies it promoted and sold. This agenda necessitated that the contraceptive industry adopt new scientifically sanctioned production processes to remain viable in the industry that the NBCA intended to dominate and direct.[13]

The NBCA consistently revised and amended the list over the course of the 1930s and 1940s, and in an effort to advertise the benevolent public health agenda of this oversight, it also broadcast the fact it was not profiting off contraceptive sales in its clinics. It typically charged between 50 and 65 per cent less for products than chemists.[14] Thus, the association was attempting to demonstrate that its scientific researches were primarily being affected to benefit public health and to guarantee the efficacy of the available technologies for contraceptive users.

The association incorporated formal contraceptive effectiveness testing into its activities in late 1934. The tests it used for this purpose were those designed and employed from the late 1920s by the BCIC. The stated aim was to undertake a 'scientific study ... of the comparative advantages of different methods [of contraception, to determine] the possible far-reaching effects of the practice as a whole'.[15] This work was inspired by the General Medical Council Pharmaceutical Committee's assertion that Britain needed to ensure 'the effective control of the quality and authenticity of ... therapeutic substances offered for sale to the public'. The pharmaceutical committee expected its definition of therapeutic substances to include 'prophylactic and diagnostic agents' and demanded that products unable to be chemically tested should be strictly 'supervised and controlled'.[16] Those claims led to the introduction of the Therapeutic Substances Act (1925), which included drug assessment legislation and penalties should a pharmacist dispense drugs of adulterated or diminished quality or potency. A comparison would be drawn against the 'presumptive legal standard' of quality as defined in two national drug manuals: the *British Pharmacopoeia* (1864) and the supplementary *British Pharmaceutical Codex* (1907).[17] Both guides specifically excluded contraceptive preparations and products, owing to the medical profession's position that contraception was not within its remit. The BCIC expressly opposed this claim and lamented that despite increased patient knowledge and demand for contraceptive prescription and fitting on medical grounds, the subject had 'hitherto not formed any part of medical education'.[18] The BCIC then set itself the task of standardising and overseeing the quality and effect of chemical contraceptives, a fact it advertised prominently and repeatedly in the *British Medical Journal* and the *Lancet* from the late 1920s to encourage professional involvement with the practice.[19]

But before activists could demand medical oversight and prescription of contraception, the issue of the physician's role regarding fertility control, independent of patient desire, had to be addressed. In 1921, British physician and birth control supporter Killick Millard used his position as Medical Officer of Health to establish the first birth control clinic in Leicester. In his official capacity he conducted a survey of 160 'gynaecologists and women doctors' to assess the profession's general stance on contraception. Of the 65 responses received, 37 were positive, with a further 14 offering 'qualified approval'.[20] He determined 'it could no longer be claimed that doctors condemned contraception', but rather, that many remained reluctant to participate owing to the current dispute about the potential physical and moral risks associated with contraceptive practice.[21] From this account it is not clear how many of the respondents were women doctors, but as Rusterholz has shown, it is likely that the significant approval received likely came from the emerging generation of female physicians who were gravitating to the profession's 'feminine areas of work' and growing openly sympathetic to the cause.[22] In 1923 *The Practitioner* dedicated an entire issue to contraception and solicited learned opinions on the practice from gynaecologists and obstetricians.[23] In 'Contraception and the Medical Profession', John Peel derided all but one of the contributions for displaying an 'obviously limited medical knowledge' of the practice, as proved by

> three assert[ions] that contraceptive usage produced sterility, a fourth alleg[ing] mental degeneration in subsequent offspring and a fifth equat[ing] birth control with masturbation as a practice 'distinctly dangerous to health'.[24]

As reports of the 'medical harmfulness of contraceptive practices' filled current medical journals,[25] along with assertions that the 'doctors oath' precluded physicians from 'reducing the population or strength of the people', the profession needed authoritative contraceptive information as much as the public.[26] The NBCA and BCIC had to address these safety and decency objections in order to convince the profession that contraception was a medical concern that physicians should understand

and employ. Luring the profession into the field would require the introduction of a contraceptive compelling medical oversight and expertise: a cheap and effective chemical contraceptive.[27] The creation or discovery of such a drug therefore became a BCIC and NBCA goal, and informed research and funding choices for three subsequent decades.

Ilana Löwy and Merrily Borell have studied early contraceptive investigations and collaborations between groups invested in the scientific legitimisation of contraception. Both agree that in the 1930s British biologists rose to prominence due to better funding opportunities and closer relationships with the local eugenics movement.[28] However, both historians neglect to address the *three* prime reasons American funding was funnelled to Britain: an existing US ban on marketing and disseminating contraceptive information care of the 1873 Comstock Act[29]; the American scientific community's reluctance to stoop to mechanical contraceptive research[30]; and the general perception that

> the British Isles are the only place in the world where work of this character can be carried on by scientists of the highest standard and the results of their work made available to the general public without serious governmental interference.[31]

Löwy charts the development of the BCIC and NBCA's early efficiency testing but credits later US-based scientists for achieving the most significant developments. She specifically cites Clarence Gamble's efforts to standardise a spermicidal testing methodology and advance contraceptive testing as a medico-legal issue.[32] Whilst this work was certainly vital to the American history of contraceptive science and technology, the desertion of the NBCA/FPA and the BCIC and their approved list for Gamble's later standardisation studies betrays a consistent historical trend to attend British research but conclude studies in the US, thus highlighting the former as the origin and the latter as the crescendo. In this vein, Adele Clarke claims that the American medical and scientific community had a quid pro quo arrangement with birth controllers not to study contraceptive chemicals and technologies until a truly medical product was developed.[33] Clarke's book argues that no similar arrangement existed in the UK, but actually the NBCA took the exact opposite position. As two very recent studies by Borge and Jones have touched on and my own published work has argued,[34] the association determined that although better and more definitively medico-scientific technologies would undoubtedly be developed, that refining and guaranteeing the currently available contraceptives, particularly the spermicidal chemicals it was prescribing and selling in its clinics, would be the best means to achieve political and medical legitimacy. A collaborative and competitive trans-Atlantic movement of ideas and a race for supremacy also informed and shaped contraceptive research.[35] Ultimately, these contraceptive movements diverged in their research aims, with the Americans focusing on developing new technologies, and the Brits overwhelmingly focused on refining and regulating the contraceptives already on the market.

A significant rivalry between the two British researchers undertaking quality and efficiency tests on spermicides provides a microcosm of this trans-Atlantic contest to define and perfect contraceptive science. Late in the 1920s, John R. Baker at Oxford University and Cecil Voge of the University of Edinburgh, were both charged with developing a standard and methodology for accrediting clinical contraceptives. Baker, a distinguished biologist, physical anthropologist and cytologist, agreed to undertake chemical contraceptive research and drug development for the BCIC and NBCA. After several years in this role, he transitioned to applied contraceptive chemistry, an untenable position in academia and so, he left for a consulting role in private industry. Voge, a recent Doctor of Chemistry, was recommended by his mentor and World Population Conference co-organiser F. A. E. Crew, to accept North American funding from the National Committee on Maternal Health and the Bureau of Social Hygiene. Both men were given identical objectives: to undertake efficacy testing of preparations exhibiting spermicidal qualities and develop a perfect contraceptive.[36] Both, working independently, developed strikingly similar laboratory techniques to assess the ability of a specific contraceptive compound to arrest and kill spermatozoa in conditions replicating the female reproductive system.[37]

These competing projects each advanced the goal of defining and overseeing contraceptive standards and demonstrated the complex professional and bureaucratic intersections between science, business and contraception during the 1920s.[38] The NBCA's funding and direction of spermicidal research resulted in the foundation of the approved list which effectively created formal contraceptive standards and elevated the association to be a respected contraceptive regulator. The pragmatic trans-Atlantic decision to engage in applied chemical contraceptive research significantly moulded the future careers of both scientists. Accepting the invitation to develop contraceptive standards marked both Baker and Voge as progressives and visionaries, but marred their academic and scientific integrity—ultimately forcing each to abandon pure research for private enterprise.

Between 1929 and 1932, Voge under Crew's guidance assessed the effectiveness and safety of chemicals displaying contraceptive qualities that were currently marketed in America, Britain, Germany and Holland. 'He tested proprietary compounds as well as the contraceptive potential of common substances (soaps, vinegar, alum, lemon juice) and showed that some chemicals assumed to be spermicidal were not'.[39] He devised a clinical method to evaluate the toxicity of each product to sperm and record how long sperm survived in a test tube.[40] He emphasised that successful spermicides must diffuse quickly and effectively in the vaginal cavity and kill or incapacitate all sperm within sixty seconds. Additionally, it should remain stable in alkaline and acidic environments[41] and be 'nontoxic, nonstaining, cheap, tolerant of storage under diverse environmental conditions [and] easy to use'.[42] Voge also studied if chemical compounds were 'lethal to gonococcus and spirochete' to protect against venereal disease as well as pregnancy.[43] He published his results testing 162 proprietary products in *The Chemistry and Physics of Contraceptives* (1933). This publication recorded the first 'extensive study of spermicides in which human sperm was employed', and it described the method and results of the three tests Voge designed and implemented to assess and rank the efficacy of spermicides.[44] The first test was a 'rapid survey'. It saw four loopfulls of sperm transferred to a slide from an inoculation loop, followed by the same amount of the solution being tested. Within the allocated fifteen-minute test time frame, this method satisfactorily assessed the general spermicidal ability of each chemical. The second test was more extensive. It involved the collection and mixture of sperm and the spermicide in a capillary pipette. The method required precise quantities of seminal fluid and a spermicide or 'diluting agent', to be sucked into a pipette with air bubbles in between to ensure the liquids mixed only when they reached the top. The solution then sat for a desired time period, prior to examination. Voge's final method held the sperm in Wasserman tubes and gradually inserted the contraceptive via capillary pipette.[45] According to Voge's manuscript, each test proved satisfactory, and could successfully assess contraceptive efficacy until 'the medical profession will aid us in ascertaining the clinical value of the preparations which are now used'.[46]

Voge betrayed his contraceptive sympathies by noting that his agenda was ultimately wider than simply understanding and analysing chemical contraceptives in the statement 'so many have refused to consider this subject seriously... [but] once the laboratory worker has systematically surveyed the ground ... then medical practitioners can apply this knowledge'.[47] In this vein he praised his competitor Baker's 'excellent technique for the examination of spermicidal substances', as, in his estimation, they were essentially collaborating towards a socio-political goal: forcing the physician's hand by proving contraceptive products to be scientifically sound.[48] However, this research made very little impact outside the greatly enthusiastic contraceptive lobby.[49]

In contrast, Baker did not consider his work was overtly political, and as a result focused his spermicidal analyses on the task of developing an 'ideal chemical contraceptive'.[50] His initial mandate from the BCIC was to investigate the 'susceptibility of sperms to poisons, including those commonly used in chemical contraceptives'.[51] He first concentrated on 'pure substances', but quickly included 'different substances ... on the market ... known to have germicidal power'.[52] Baker's tests were developed using cavy rather than human semen, as it was cheaper and easier to access a 'perfectly fresh supply'.[53] Baker's 1929 publication of his pessary test method utilised animal sperm, however within two years tests employing human sperm were being prioritised. In the July 1937 publication of his official contraceptive testing methods in the *Journal of Hygiene*, Baker clarified 'our tests are nowadays [all] done with human semen' owing to differences in killing power between the two.[54] This transition prompted the NBCA to posit that some spermicidal failures were the result of naturally occurring differences in semen potency. Understanding the 'standard of ordinary semen' would have a definitive impact on efficacy testing as it meant contraceptive potency tests could effectively be targeted towards sperm with higher resistance, thus ensuring a higher margin of efficiency. However, no evidence remains regarding the pursuit of this enquiry or its findings.

Baker's spermicide test required the preparation of two batches of neutralized 'buffered glucose saline', a neutral substance replicating vaginal conditions.[55] One batch was a control, and the other was used to test the spermicidal capacity of substances. Both batches sat in a damp chamber, until they reached body temperature. The chemical being tested was then dissolved in 0.9 per cent saline, at various concentrations, as Baker aimed to discover the lowest concentration of each product lethal to sperm within thirty minutes. The active agent was applied to the solution and fifteen minutes later the sperm. Air was introduced at fifteen, twenty-five- and thirty-minute intervals to prevent the sperm from suffocation. The slides were then examined microscopically and sperm motility applied a number: III, II, I or 0. III indicated high motility, and 0 indicated none, with + added if movement fell between classes.[56] If the control tube motility was III or III+, the test results were recorded and repeated thrice to ensure the outcome.[57] The lowest concentration sufficing was called the 'kill concentration'. All subsequent tests were then conducted at half the kill concentration of a product to rank spermicidal ability by assessing sperm motility at half strength.[58]

As mentioned, both ventures also intended to develop a contraceptive that was superior to those currently available. The more experienced Baker consistently worked towards this objective, whilst Voge diligently undertook each task methodically. Both ultimately agreed that most contemporary chemical contraceptives had 'limited spermicidal power', thus necessitating a more effective contraceptive, but neither successfully developed one during his initial tenure.

As Chapter 2 described, Voge and the North Kensington medical committee collaborated with contraceptive manufacturer Gilmont Products to develop and test a contraceptive cap that met the requirements and desires of that site's physicians. However, the collaboration also included the formulation of a spermicidal foaming jelly, PermFoam, which was dubbed effective by association therapeutic and research experts Joan Malleson and Gladys Cox in their comparative analyses of contraceptive technologies[59]; but even Voge admitted it was prone to fail in an insufficiently moist vagina, or disintegrated if it came into contact with moisture during the manufacturing process or storage.[60] The NBCA did in fact test Permfoam as a stand-alone contraceptive in 1935, but the trial was halted after a spate of pregnancies. The product was condemned by two clinics and two private doctors as '(a) not reliable (b) messy [and] (c) irritating in use'.[61] Hence, Voge's subsequent focus on rubber contraceptives. Once the scientifically sanctioned and standardised GP cap was guaranteed by Voge's rubber tests, it became the NKWWC's primary contraceptive cap.[62] However, neither product was ever acknowledged as meeting the requirements of Voge's American contract to create a superior product, rather both were ultimately imitations of currently available products—one inferior and the other made-to-order.

Ultimately, this failure devastated Voge's academic career. His American financiers, on his mentor Crew's advice, dubbed his failure to develop 'a new and better spermicide' a contractual breach and rescinded his funding.[63] It was further claimed that Voge had become too sympathetic to contraception, a position that compromised his ability to conduct unbiased scientific research.[64] Crew scathingly dubbed Voge a 'traitor to science' for flirting with 'applied science' by collaborating with a manufacturer to develop a contraceptive product. Crew guaranteed Voge's dismissal by refusing to supervise further research.[65]

With this discharge, the supremacy and authority of Baker's contraceptive testing method and results was guaranteed. In 1935, he formalised the authority of the 'Baker test' for evaluating contraceptive efficacy in *The Chemical Control of Contraception*. Despite it being printed two years after Voge disseminated his method, Baker's was deemed more authoritative by the NBCA which was already utilising his method to inform and direct the merchandise and methods its clinics prescribed. Baker's success was not without consequence. His unsanctioned foray into spermicidal research was deemed to be too overtly applicable to the contraceptive cause and he lost his position in the Oxford Department of Zoology.[66] But it was sufficiently independent to encourage the NBCA to direct further funds and support to his research and development agenda. The additional guarantee of a laboratory at the Oxford Department of Pathology kept Baker dedicated to the contraceptive project.

For the remainder of the 1930s, Great Britain was the epicentre of clinical contraceptive testing. All funding, even from the National Committee on Maternal Health and Bureau of Social Hygiene in the United States, was funnelled into Britain to ensure Baker's contraceptive efficacy and acceptability testing could continue unfettered.[67] Despite this, there remained a professional stigma within the sciences regarding applied research that was contemptuous of this controversial emerging field of study. Thus, the birth control lobby alone had to fund and support this work, as no independent research bodies, hospitals or universities would.[68]

From 1937, Baker's method became 'used as a standard test by the [NBCA/]FPA' to ascertain the spermicidal efficiency of chemical contraceptives included or excluded from their annual approved list of contraceptives. The association consistently cited that this method offered 'reliable and reproducible results within the limits of individual variations of semen and homogeneity of the preparations used' to reliably demonstrate 'whether a preparation contains an adequate amount of spermicide or not'.[69] This standard remained into the 1950s, and during discussions with the British Pharmacopoeia Commission about transferring oversight of chemical contraceptives to the body, it was submitted as an appropriate procedure for establishing efficacy.[70]

Whilst Baker was perfecting his method, the NBCA and BCIC spearheaded and oversaw a pre-1937 application of the test to assess and approve contraceptives prescribed in association clinics. Just as Löwy has demonstrated Clarence Gamble tried to do in America, the association intended to 'put pressure on commercial manufacturers of contraceptives to improve their products'.[71] The difference was that the British approach would not strongarm manufacturers, but would rather encourage the 'makers of various commercial pessaries' to cooperate with Baker's research.[72] A collaborative sensibility is borne out in the minutes of a 1935 meeting between Baker and M. V. Bowler, the Chief Chemist at Gilmont Products Limited. Bowler initiated the meeting to discover why the NBCA had determined 'G.P. Solubles were not so effective as others with which comparison was made', and demanded to know 'why Dr Baker's tests gave such different results as their own'.[73] Two important points can be gleaned from this record: that many manufacturers had provided Baker samples of proprietary contraceptive compounds for testing and that Gilmont (and presumably other manufacturers) was conducting its own tests to achieve some generally accepted standard of contraceptive efficacy. This proves that commercial and benevolent societies had separately and jointly attempted to scientifically legitimise birth control far earlier than was previously thought, and that attempts to standardise contraceptive technologies through an alliance with clinical science had already been effected by individual manufacturers—it was just the formalisation of a universal standard that had been defined by Baker through the approved list.

The meeting report explained that Messers Gilmont 'had carried out considerable experimental research work for hospitals, clinics and doctors' and that these interactions had resulted in significant product revisions. Bowler insisted that when Gilmont's products were consistently 'not giving satisfaction' in acceptability or result, whole new preparations were evolved. This comment reveals an aspect of contraceptive testing neither the NBCA, BCIC, nor their American counterparts attended before 1935: the acceptability of contraceptives for users. Tolerability would eventually become a major focus of contraceptive use and product loyalty.[74] Gilmont criticised what it deemed an unfair appraisal of its spermicide. It complained that Baker's test examined compounds as independent contraceptive products and failed to consider the manufacturers' advised method of use. GP Solubles, it justified, was designed and prescribed as a complement to a barrier appliance and was never to be used as a stand-alone contraceptive.[75] This concern may seem minor, especially as the NBCA/FPA's official position was that 'reliance should be placed on a combination of methods and not on any single one'.[76] However, as knowledge and acceptance of contraceptive methods and appliances grew, along with the ascendance of the association and its clinics, pressure on contraceptive manufacturers wishing to have their products administered and sold at those sites to meet Baker's standards increased.[77] Gilmont had proved its dedication to the ideal of product testing and regulation by soliciting and meeting the requirements of clinics, hospitals and other groups, 'prior to the time of putting [a product] on the market'.[78] It was therefore happy to cooperate with the principle of an association-led effort to standardise and regulate contraceptives, it only queried the NBCA's chosen method. Bowler explained that Baker's findings had caused 'grave disturbance... in the minds of [Gilmont] Directors', and they particularly dreaded his results might be published before they could challenge them with their own evidence.[79]

The company appreciated and even agreed with the association's attempt to standardise contraceptive testing, but formally and crossly disagreed with Baker's technique. Gilmont criticised the use of cavy semen as flawed and '[in]comparable with ... tests employing fresh human semen' and further alleged any attempts to 'reproduce coital secretions' were impossible as all individuals 'possesses a different protein concentration'.[80] Gilmont pitted its science against Baker's to discredit his techniques. After 1938, Gilmont extended its criticism to his ethics, when Baker and manufacturer British Drug House announced the development of a completely new spermicide: VOLPAR gel and paste, abbreviated from voluntary parenthood.[81] The NBCA financed Volpar, a spermicide supposedly 'superior to all other chemical contraceptives',[82] without informing other manufacturers who had delivered products and unguarded formulas to Baker to facilitate the association's testing programme.[83] Contributing manufacturers took umbrage. They had agreed to work with the association in a regulatory context but protested an act of 'industrial espionage disguised as product standardisation'.[84]

As was discussed in Chapter 2, the issue for Gilmont was that from the late 1920s it had collaborated with Voge's Scottish research and applied his efficacy tests and standards to its products.[85] The company explained that 'Edinburgh University and [its] own laboratories' had successfully developed a new contraceptive which could theoretically be used alone and had been tested extensively and proven effective by gynaecologists in hospitals and clinics. The product was given the 'trade name of PERMFOAM'.[86]

In Baker's terse response to Gilmont's complaint, he insisted his tests were superior to Voge's, and derided the company for having used and continuing to support an inferior testing technique.[87] By allying with Voge, who was initially an equally viable candidate to become the eminent contraceptive scientist, the medico-scientific integrity of Gilmont's products and testing was dented when Baker's scientific method won out and he developed Volpar with its rival.[88] Gilmont alleged that Permfoam and its later products GP ointment and soluables were sidelined by the NBCA, in favour of Baker's product.[89] It further claimed the repeated failure of these products using the 'Baker test' constituted bias and demanded an impartial re-examination.

There was no objective way to resolve this dilemma as the NBCA, an organisation highly invested in the scientific legitimisation of contraception, had assumed the duty to regulate and oversee all technologies of the practice. 'The simultaneous development of two methods for producing or testing a scientific or technological product is not unusual' in the history of discipline formation and contestation.[90] Graeme Gooday explains Voge and Baker's race to become the expert and authority on contraceptive science was typical, but that the associations intervention and arbitration was a questionable use of its regulatory authority, as its relationship to Baker and his method was closely related to its political and financial success.[91]

Gilmont was unable to satisfactorily resolve this issue with the NBCA-led contraceptive testing method. It remained openly hostile towards the association until 1942 when its researchers developed a soluble pessary that Gilmont deemed superior to Volpar and wished to sell in NBCA clinics.[92] The association was unsurprisingly 'extremely interested … in a product of this nature' and offered to clinically test it. Gilmont agreed, as long as Baker nor any of his associates would be involved in the trial. The NBCA agreed that 'spermicidal testing would be carried out by a group none of whom were connected in any way with the old Dr.

Baker Group'.[93] Despite this dispute, in the intervening years Gilmont maintained annual testing to keep its products on the approved list, and guarantee NBCA clinics and patients loyal to its contraceptives had unfettered access.[94]

Voge, unlike Gilmont, refused to quietly resolve his dispute with the NBCA. He had gambled his academic career on defining scientific standards for contraceptive technologies and had been professionally sidelined by the association and its competing study. When Volpar was revealed in 1938,[95] Voge contacted Theodore Fox the editor of the *Lancet* to expose that the NBCA held a 'financial interest in Volpar'. He alleged that the association's product endorsement on the grounds that its efficacy was 'proved by laboratory and clinical tests' was suspect as it had paid its own researcher to devise and undertake those tests.[96] Voge sent a 'similar' letter to *Manufacturing Chemist* stating that 'the NBCA could no longer be considered an impartial body so far as the testing of contraceptives was concerned'.[97] Both Baker and the NBCA responded in the latter journal. The association was in the clear as it had no commercial involvement with the product having refused distribution rights when they were offered in favour of free supplies[98]; Baker addressed Voge's 'technical criticisms' and the association publicly asserted that it 'had no financial interest in Volpar' and that it was an impartial body ready to consider contraceptives produced by any firm.[99] Löwy has shown this latter claim to be suspect, as, when Baker and British Drug House attempted to advertise Volpar in North America, the National Committee on Maternal Health which had diverted funds to Baker's studies after sacking Voge, raised serious concerns.[100] Executive Secretary Raymond Squire was damning in his correspondence asserting'it has not been [the committee's] policy...to supplement commercial subsidy or even finance research where there is a prospect of such financing being done in toto by a commercial organisation'.[101] Although the committee didn't suspend Baker's grant after this ethical violation, it did specify that its funds could not be used to support any investigations made in connection with private industry.[102] The issue quickly disappeared after this admonishment, and all accusations of impropriety, both Baker's plagiarism claims and Voge's retaliatory assertions of corruption and collusion were terminated.[103] Once the resultant 'difficulties with other firms on the approved list had died down', tests were resumed and the association took steps to formalise itself as the British regulator of contraceptives. Despite the spectre of doubt regarding the partiality of the association, its list never lost the support of manufacturers who saw value in achieving accreditation for their products from the primary national purchaser and distributor of contraceptives.[104]

During the 1930s, there was an established value in scientifically examining and standardising contraceptive devices and compounds in both Britain and the United States. But a consistent transnational difficulty was found in recruiting 'first class young men to take on research work which had birth control as its object'.[105] This was largely due to the medical and scientific profession's ongoing reluctance to engage with birth control as a social practice, despite acknowledging a theoretical 'interest in the scientific study of contraception for reasons of health'.[106]

By 1940, the association had established a regulatory niche for itself and was becoming the renowned public face of contraception in Britain. It formally constructed frameworks for overseeing products that had previously been too marginalised to warrant any formal recognition. The association was adamant its work was imperative to national and private interests and instituted annual regulatory checks to ensure contraceptive quality was maintained and improved where necessary. As is discussed in the rest of this chapter, the FPA also turned its focus to rubber standards, and harmlessness and acceptability studies, but these studies fell more broadly under the applied science banner. The association's efficacy, safety and tolerability goals became integral to 'giv[ing approval to] products, which in the opinion of [its] medical members are satisfactory' and useable for the 'guidance of purchasers'.[107] Much of this work resulted from data collection innovations initiated at the NKWWC and implemented in association clinics which alerted medical staff to the serious risks associated with serial pregnancy, and aimed to combine contraceptive science with current preventive health frameworks.

Contraceptive Data Collection and Scientific and Clinical Statistical Research

As has been shown, many of the early clinic management methods and protocols emanated from the Society for the Provision of Birth Control Clinics (SPBCC) that had been amongst the earliest established in Britain. During the 1930s the NKWWC and its staff were keen to modify and standardise the centre's practices and procedures to aid the wider contraceptive cause. That decade the clinic became the association's primary site of epidemiological investigation into women's sexual, maternal and contraceptive health. It pioneered methods of data collection, management and analysis that bolstered and publicised the necessity, as well as the safety and efficacy of current and emerging contraceptive technologies.

Helena Wright, the centre's Chief Medical Officer from 1928 until the 1960s, dominated and directed its form and function, and in some instances also set the association's medico-scientific aspirations. Owing to an arrangement negotiated with Margery Spring Rice before accepting the role, Wright 'had a free hand to make whatever changes she wished' and she efficiently medically managed the clinic during her tenure.[108] She and Rice shared the BCIC, and later association's, conviction that contraceptive advocates must

> investigate the sociological and medical principles of contraception; the possible effects of the practice upon physical and mental health; and the merits and demerits of all possible methods.[109]

This was in part a tool for legitimising contraception for social and political ends, and also a means to alleviate potential concerns about the safety, efficacy and ease of use of currently available contraceptive methods. As Kate Fisher has demonstrated through oral histories, patients experienced 'great anxiety' regarding contraceptive devices and methods despite clinicians guaranteeing that prescribed products and methods were safe and effective.[110]

To that end, Wright and her BCIC, NBCA and NKWWC colleagues undertook significant research into the best means of collecting and corralling patient medical and sociological data in order for it to be medically and scientifically useful. Wright was, from 1929, acquainted with contraceptive research and development owing to her employment as a scientific researcher for the BCIC. She was then charged with undertaking two investigations, one pure study examining the 'chemistry of the female genital tract' and a statistical assessment of the effectiveness and safety of the Gräfenberg ring intra-uterine device.[111] In addition to standardising clinic and medical protocols and treatments, Wright also pushed the association to focus on standardising patient and efficacy data collection and management via clinic case cards which became a valuable scientific tool for legitimising contraception as a practice and a therapy. Achieving this socio-political goal would have been impossible without the uniform design, completion and maintenance of patient records at SPBCC sites; and their gradual export to all association clinics over the subsequent decades. The patient case card and the data it provided were quickly identified as an effective means of proving the usefulness and effectiveness of contraception as a preventive health tool. The NKWWC under Wright's direction also acceded to design and undertake applied clinical trials of proprietary contraceptives, as the statistics that were derived from this work would have the potential to improve individual contraceptive products, and also to facilitate the advance of contraceptive guidance and techniques by medical staff.

The decision to collect potentially useful data was not isolated to the NKWWC; it became a standard practice in contraceptive clinics during the mid-1920s, as the importance of providing personalised and continuing care for gynaecological and contraceptive patients became generally accepted.[112] The initial case cards utilised at the NKWWC were based on those implemented at the Walworth Clinic. That site was opened in Elephant and Castle in 1921 and was under the medical direction of physician and birth control advocate Norman Haire. The early case cards collected important medical and sociological data: marriage length, parties current age, husband's occupation and weekly income, how patients learned of the clinic, and a full fertility history—detailing the number of pregnancies, live births, miscarriages (including abortions), as well as live children and total losses.[113] These initial cards proved sufficient until the mid-1930s.

Both the BCIC/SPBCC and Marie Stopes claim credit for pioneering the recording of birth control patient data on a uniform card.[114] Edith Howe-Martyn and Mary Breed's *The Birth Control Movement in England* (1930) attributed the practice to Stopes; in it they explained that Stopes 'publish[ed] an interesting report ... of the first 5,000 cases dealt with' in her Mother's Clinic for Constructive Birth Control within four years of its founding, and this was achieved 'by keeping case sheets with records of many data'.[115] So, just as Stopes' was the first British clinic, she was also the first to record and publish a statistical analysis of patient data in February 1925.[116] In *Equipping a Birth Control Clinic* (1934), Stopes clarified 'naturally, ... I was first to initiate [case sheets]', and that as their pioneer, she designed the cards and determined their appropriate contents. Stopes restricted the data collected to cover childbearing and marital history to 'ascertain all the facts a scientist might wish for without thwarting the objects of the clinic', and ascribed little value to data outside of immediate clinic use.[117] Stopes stipulated the importance of including the contraceptive appliance prescribed and its 'exact size' on each case sheet for reference, renewals and to enable future comparisons, should childbearing alter the patient's reproductive physiology.[118] Of course, this specific data had the added benefit of validating her claim that her preferred contraceptive device, the Pro-Race cap, later renamed and trade-marked as the Racial cap, was the 'best type of vaginal, all rubber cap' and that this was proved by patient records from her clinics.[119] This declaration was made amid her disagreement with Haire about the ideal cap. Stopes' clinic data (as well as that allegedly collected from the 'Walworth School' of clinics) was cited to denigrate Haire's preferred Dutch cap and promote her own Racial articles.[120]

The First Five Thousand, whilst purporting to be a statistical analysis of the 'careful[ly maintained] records of every case actually examined and assisted', offered little substantial analysis.[121] Rather, it justified the necessity of contraception, through some numerical deduction and self-promotion. The text did evidence a negative relationship between female fertility, number of pregnancies and the likelihood of miscarriage or infant loss. Stopes further demonstrated that 'repeated pregnancies [we]re in themselves a cause of infant mortality which no environmental conditions can counteract completely'.[122] This topic was so novel, that later investigators credited her for statistically demonstrating the link. Norman Himes, who published the first statistical analysis of the NKWWC's patient data in 1929, utilised some of Stopes' fertility statistics to argue that 'both sets of figures illustrate the same principle: the tendency for the rate of loss to surge with increased frequency of pregnancy'.[123] The social relevance of this data was later bolstered by the sociological findings of the Departmental Committee of Maternal Mortality and Morbidity which supported both Stopes and Himes' claims in its interim and final reports. Contraception undoubtedly had an important application as a preventive health tool for women for whom pregnancy could be risky.[124]

The first assessment of the case cards devised and completed at the NKWWC was 'British Birth Control Clinics: Some Results and Eugenic Aspects of their Work' (1929) by Norman Himes.[125] This article utilised North Kensington data, along with that of eight other early British contraceptive clinics. This investigation used 1000 NKWWC case cards to draw conclusions about clinic attendees, their social class and husband's occupational status, typical weekly wage, average parental age, duration of marriage, family size, typical fertility and miscarriage rates and the correlations and contradictions between them. The collection of such detailed sociological data was then novel to SPBCC-affiliated sites. Stopes was reticent to collect non-medical information, as she worried the perception she had a political design might impact the collection of an honest and complete medical history.[126] The medical staff at North Kensington and its affiliates had no such qualms; there the political agenda was more seamlessly assimilated into the site's overall medical plans and programs.

Himes' article included a small statistical inquiry that related to patient's historic use of contraceptives to limit offspring. However, this data was exclusively collected from Liverpool clinic case cards and only considered 109 total patients; 33 (30.3 per cent) of whom fell under the category 'no use of contraceptives mentioned'.[127] The NKWWC management appreciated this question could produce valuable data about the acceptability and effectiveness of contraceptives and pursued this in subsequent iterations of the case card.[128]

The most innovative aspect of Himes work was comparing the 'occupational status of husbands' and 'economic and social data' of British and American contraceptive users. The latter (sampled from New York and Chicago) typically reported approximately double the weekly income of their British counterparts. Himes understood this resulted from the strict conditions in America which limited the provision of contraceptive information to those 'who need ... advice for medical reasons'.[129] The groups targeted by American and British contraceptive sponsors were significantly distinctive and in the UK contraceptive advocates established clinics in poor areas with major populations of disadvantaged and uneducated clientele. There the provision of contraception was both a benevolent act and a means to limit the propagation of 'delicate, diseased or unwholesome infants which [we]re unlikely to live and therefore a great source of expense and waste'.[130] The connections between eugenics and birth control were more openly admitted in the pre-NBCA social welfare era, than after its establishment when its work was actively promoted as being based on contraceptive science.

British clinic management assumed that better bred and educated citizens would gain contraceptive advice and instructions from other sources, primarily private doctors.[131] However, American birth control clinics served a broader clientele. Himes identified that 60 per cent of attendees in Chicago were of skilled, professional or managerial classes, and in New York, over 20 per cent of the 1,000 cases investigated made over $50 weekly—placing them financially amongst the middle and upper classes.[132]

By comparison, the North Kensington clinic, and most of those the association founded, primarily serviced the unskilled and semi-skilled. As that clinic was located near a 'London bus route' terminus, a full nine per cent of the total clientele reported their husbands were 'bus conductors or drivers, motor mechanics and chauffers'; this figure combined with a miscellaneous 23.8 per cent that was 'skilled or semi-skilled' and a further 37.9 that were 'definitely unskilled', to create the bulk of the local clients. Those falling into the 'higher business and professional' sets comprised a mere 17, or 1.7 per cent of the first 1000 local cases.[133]

The article resolved that the available data could not be inferred into any definite calculation of the contraceptive competence or aptitude of British and American users. US clinics clearly reached members of a superior economic cohort, from which a correlation between income and intelligence could potentially be assumed; however, Himes determined that any such assertions would demand 'a great deal more fact-collecting and fact-analysis [than] has been carried on'.[134] Thus, demanding more data and statistical investigation.

That same year, Norman and his wife Vera Himes published 'Birth Control for the British Working Classes: a Study of the First Thousand Cases to Visit an English Birth Control Clinic'. This was exclusively based on NKWWC clinic records that were used to make a more substantial local sociological analysis. The investigation defined the typical North Kensington patient, her fertility, gynaecological and marital history, class and economic status, capacity to learn and apply contraceptive technologies and attempted to assess the effectiveness of those technologies in actual use.[135] The breadth of data that was collected at North Kensington 'reflected [the efforts of] a new research committee' working in cooperation with the BCIC that set 'the investigation of methods and the endeavour to improve Birth Control technique' as its primary goal.[136] This effort formed the model for the FPA's future work to standardise sociological research methods at its clinics.

The article identified the two primary subjects of applied scientific investigation the association pursued using patient data collected and maintained on cards—contraceptive acceptability and success in actual use. The Himes's study flagged the importance of comprehensive follow up visits (in patients' homes, if needed) to assess the actual efficacy of contraceptives in use and the uniform recording of the results to develop and guarantee future contraceptive prescription and instruction. In 1927, after examining NKWWC records, Vera Himes was shocked to discover how few women returned for follow-up appointments. She claimed this meant that 'the clinic had no way of ascertaining the value of its work' and that any statistical study published without acknowledging this inadequacy was redundant.[137] This was also a thinly veiled attack on Stopes' sociological research method that assumed women who failed to return did so due to an absolute satisfaction with her prescription; rather than a dissatisfaction with the invasive, messy and time-consuming two-part cap and spermicide prescription that many subsequent historians have asserted was frequently deemed unsatisfactory and abandoned.[138] To address the early observation of this failing, Himes selected 96 women who had visited North Kensington within the last two years to determine 'whether the patient was using the appliance or not; if not using it, the reasons; if using it, whether or not it was satisfactory'.[139] Himes tracked down 68 of these women, only 58 were even fitted with an occlusive contraceptive.

> Of this number, 26, 45 per cent., were not using the appliance recommended while 32, 55 per cent., were employing the device, 18, 31 per cent., satisfactorily, and 14, 24 per cent., unsatisfactorily. In this last group all but one woman had become pregnant.[140]

These findings fit especially well with Kate Fisher's oral history studies which support the SPBCC findings that many users (and their husbands) so disliked these methods that they either trialed and interchangeably used a variety of contraceptive methods over the course of their sexual lives, or ultimately returned to coitus interruptus and 'appear quite satisfied'.[141] The Himes' further asserted that these follow up visits proved that it was 'absolutely indispensable to the intelligent service of the clinic' to maintain contact with and assess patients' experiences and satisfaction with the contraceptives, primarily caps, they were prescribed. The two statistical investigations that the pair compiled using the spartan resources and data amassed from the early NKWWC records, paved the way for future studies, many of which directly replicated or followed on from the issues raised by the pair.[142]

These early investigations were made using the original case cards that were developed in the mid-1920s as the NKWWC established itself. After 1927, in conjunction with the BCIC and in 'cooperation [with] private practitioners', the clinic developed and 'much amplified' its case cards that were until then reproductions of the Walworth cards.[143] This transition allowed for the broader use of this data to facilitate patient treatment and training in contraceptive clinics, and encourage social, medical and political backing and sympathy.

After the NBCA was established the NKWWC overtook Walworth to become its model clinic. In order to capitalise on official concessions regarding the acceptability of preventive contraceptive therapies laid down in the mid-1930s, the NKWWC formalised and standardised its medico-scientific structures and processes. Contemporaneously the clinic was being physically expanded through the purchase and renovation of the building next door, the launch and expansion of educational and promotional activities, and the evaluation and refinement of clinic practices.[144] Contraception was gaining socio-political and medical traction, but much legitimising work was still required, and the NBCA and NKWWC medical committees moved to adapt the clinic's operations and practices directly to that task.

In 1934 the Chairman's Report of the NKWWC medical committee announced the necessity of a detailed revision' of current patient record keeping. The report confirmed the centre was updating case cards 'in order that greater efficiency and uniformity of records may be maintained'.[145] This decision required a formalisation and standardisation of the site's data collection and completion processes; and to guarantee the veracity of its data, mandated 'a standard form of [case card] entry' signed off with doctor's initials to prevent future medical staff using the cards from having to 'draw conclusions (possibly erroneous)' from imprecise records.[146] It was agreed that collecting and maintaining certain specific sociological and medical data was imperative to the association's applied scientific aspirations and to the process of standardising and tailoring medical services to its national clinics. So, North Kensington became the test site for collecting and applying this data. In addition to the previously defined medical, economic and family data, cards were redesigned to determine where patients 'heard of the clinic', who they were 'referred by', as well as to record information about their last monthly period and to fully incorporate a patient history of contraceptive effectiveness and failures. Further, 'conspicuous... red star stamp[s]' were designed and ordered to facilitate the recording of information about new/changes to contraceptive prescriptions.[147] Other new stamps would facilitate the collection of specific health data at the six-month check-up and the recording of gynaecological conditions after birth.[148] Thus, the association envisioned and actively worked to ensure its cards became a powerful tool for recording 'clinically' and 'sociologically' useful evidence to comprehend, guarantee and advertise the efficiency and sufficiency of its prescribed products and methods in actual use amongst class-, education- and region-specific and national cohorts.[149]

The development of data collection to incorporate non-medical information also required in-clinic data collection processes to be regimented and supervised. At each North Kensington session patients already underwent two assessments: both lay and medical. This was standardised and worked effectively; but the revision and extension of the data collection duties, demanded complete uniformity in case card management and completion between lay and medical staff. Addressing this need led to the definition of clinical protocols to systematise card language and check completion to ensure the utility of data for future investigations.[150]

Before 1935 the NKWWC lay staff ran the waiting rooms, timetabled doctors' visits and collected and recorded 'essential non-medical particulars' from patients. In addition to being responsible for maintaining correct referencing and filing, these volunteers undertook some statistical data collection and compilation duties in service of the association's political and medical efforts to professionalise itself and its contraceptive services.[151] Without this administrative and analytical work, the early sociological data collected on the cards would have been useless. However, despite their contributions and efficiency, following the 1935 revision, the medical committee resolved that medical officers and nurses must have exclusive responsibility for collecting and recording standardised patient data, as ultimately all of this information was medical, but also had a sociological value. The work of statistically assessing and extrapolating valuable data from the cards had already become too big and important a task to be left to untrained women, and although the association was keen to stress the important contributions these women made to its applied scientific goals, it had expanded so much that 'trained help' was necessitated to effectively pursue this agenda.[152]

The June 1934 NKWWC medical committee meeting minutes included firm new rules and methods for case card completion during the course of a typical doctor-patient interaction. It began with a general evaluation of the patient's family and lifestyle circumstances and was followed by a series of physiological and fertility questions to compile a full health history. Details of the patient's last monthly period including her approximate volume of blood loss and history of menstrual pain were deemed imperative to understanding her overall health, as well as her fertility. Further discrete and collective sexual and gynaecological insights could be ascertained if patients reported and qualified a history of painful dysmenorrhoea and dyspareunia. This was followed by physical vaginal examination by hand and with speculum. The doctor recorded the size, tone and state of the vaginal opening, wall and outlet, their apparent health and any discernible erosion, and for mother's, noted the scale of perineal tear. To further ascertain reproductive health, 'the position and size of the uterus was recorded, along with any apparent ovarian or fallopian tube abnormality' along with any 'disabilities from frequent pregnancies', to 'specifically note' and advocate for women for whom 'medical evidence was found for the avoidance of pregnancy'. Finally, contraceptive methods, previous, current and prescribed, were detailed, along with specific instructions for use. 'All prescription notes were to be signed with [the doctor's] initials' to ensure accountability.[153] This case card completion standard became the NKWWC and later the NBCA's clinical data collection method to achieve their dual goals of being able to both guarantee apt contraceptive prescription and amass useful data.

The case card that the NBCA sought to export and apply across its growing clinic network was based on the NKWWC model. This effort began in the mid-1930s, at that time the association's medical sub-committee first surveyed member clinics regarding the local implementation of record collection methods that had been refined at North Kensington and Walworth. As part of its pitch, the association promised that broader data collection from further afield would facilitate larger scale and scope statistical surveys and results, in addition to streamlining practical clinical processes and clinic functions. However, there was not initially a huge enthusiasm and in 1937, only 27 of the 54 NBCA affiliated clinics employed NBCA-led case cards and data maintenance methods. In order to accommodate the financial and operational needs of individual clinics, the committee recommended those clinics not yet using it 'consider adopting it when their present supply was exhausted'. This position did not sufficiently comprehend or address the causes of this reticence, many clinics, including Birmingham, specifically objected to the standardised card on the grounds that they were not fit for purpose, specifically that the card 'contained too many extra questions'.[154] The NBCA's response was to disseminate ' notes on the filling out of case cards'; however, no evidence exists that these instructions included the NKWWC's abbreviation schedule and thus likely did not fully address these concerns.[155] Ongoing non-commitment to this agenda prompted educational efforts to support case card usefulness for applied scientific investigations. On behalf of the association, North Kensington specialist Mary Macaulay, along with Margaret Jackson, the physician who founded the Exeter clinic, introduced programs on proper 'Record Keeping' and management and for publishing a 'periodical report' using the association's patient data.[156] Making the data overtly useful was a means of encouraging clinics to embrace the technology as well as cede operational autonomy to the centralised association management. As Rusterholz has identified there was at this time a far-reaching effort, primarily effected by female doctors, to be able to rely on and cite this data to evidence the efficacy, safety and success of contemporary contraceptive efforts, and it was ultimately this utility that overcame objections to the association-wide case card.[157]

Once case cards became standard in the post-war period, various specialist clinic versions were advanced for gynaecology, marriage difficulty and sub-fertility, for failures, and finally for training medical students.[158] This expansion vastly increased the number of people completing cards in now notably busier clinics, and demanded policies to guarantee consistent completion and maintenance. The FPA executive and medical committee directed that clinic doctors were to oversee and guarantee each site's sociological data collection programs by modelling, teaching and overseeing junior staff and trainees in the broad importance of true and full patient records.[159] Despite these efforts, the problem of uniform card completion extended into 1960, when it became apparent that some doctors were not meeting the defined standards and were thus risking the overall accuracy of the records as well as their potential to serve a broader sociological function.[160]

Early in the 1930s as the NBCA began to ramp up its laboratory-based research activities, a 'periodic rounding up of [clinic, patient and contraceptive] statistics' was proposed as a sociological equivalent.[161] The recent Himes' reports proved this data was worthy of publication and would be important for maintaining and improving clinical practice and contraceptive efficacy and acceptability in actual use. It was hoped that such studies would be acceptable for the *British Medical Journal*, but the ever pragmatic association was prepared to publish in *Eugenics Review* in order to guarantee publicity for this branch of its scientific work.[162] To effect this goal the NKWWC fixed on a scheme that would employ a layperson, to analyse its patient records and draw sociological conclusions that might be of interest to Sir George Newman the Chief Medical Officer of the Ministry of Health who was a renowned supporter of gynaecological clinics and an advocate of the use of statistics for the promotion of public health programs.[163] In 1935, Mrs Wylson a clinic volunteer commenced an analysis of data from '1,000 cases at the Birth Control Clinic', and '250 cases at the Gynaecological Clinic'.[164] Upon its completion in May the committee intended to submit the report; however, this never eventuated owing to Newman's abrupt retirement, and replacement of the far less sympathetic Sir Arthur MacNulty.[165] Despite this setback, Mrs Wylson's work was still deemed valuable and she was appointed Clinic Superintendent and commenced biannual reporting on patient attendance and collecting data to study patient's failing to return.[166] These two issues were intimately linked to the success of the clinic and the medicalisation of contraceptive practice more broadly. This utility gave these analyses scientific and political resonance.

Late 1937, former NKWWC and current NBCA medical committee member, Carlos Blacker promoted a further extension of the collected sociological data to incorporate religion and income into future statistical investigations. Blacker, ever the savvy scientific propagandist, lobbied all 52 association clinics regarding potential research avenues. Specifically,

(a) The possibility of collecting information on declining fertility for the Population Investigation Committee; [and]
(b) The careful recording of the number of patients lost trace of.[167]

This attempt marked the first large scale, association-wide effort to co-ordinate case card data to answer specific research questions. Through its definition of a prospective investigative agenda that applied its patient data, the association recognised an opportunity to produce scientific statistics relating to various aspects of British contraceptive opinion, use and efficacy for government and medical authorities' consideration. However, before any potential results could be disseminated, the method had to be trialled. For this the NBCA turned to its long-term affiliate, the Eugenics Society, as it was experienced in pioneering and undertaking sociological and social welfare focused statistical investigations through its work on family history and heredity. That group, still actively funding much of the NBCA's work, agreed to allocate funds to 'employ a worker for six months to visit clinics and extract [and compile] the required information from the records'. Following the success of this small study, the NBCA undertook the work internally.[168]

A flurry of clinical and statistical investigations using the current case cards and developing new versions to collect specific sets of data followed. The most significant statistical investigations that the association concentrated on assessed patient satisfaction, the simplicity and efficiency of contraceptives, and failures. The NKWWC and NBCA were each anxious to use these findings to benefit clinic patients and their reputation and authority.

Just as Vera Himes had previously identified, both groups observed great potential value in studying clinic failures; both patients who failed to return for follow up treatment, fittings or prescription renewal, and of method or product. These two lines of inquiry were delineated and used to 'determine whether or not the client has so learned the technique taught her that she can carry it out herself' and also the acceptability and integrity of prescribed products and methods.[169] This work began in earnest in April 1937 when Wylson concluded an analysis of 2000 case cards to determine failures to return to the NKWWC. She found that forty per cent of patients on cards 6000–7000 had failed to return for follow up services, but that the number fell to only eight per cent on cards 7000–8000.[170] Of the latter, approximately seventy per cent were referred by an external medical authority, and a further 10 per cent were as yet unmarried, indicating that patient trust and adherence to the prescribed contraceptive practice was higher when patient's exposure to the clinic was obtained through a medical practitioner or came before her marital sex life began. A more basic assessment that considered total clinic attendance in six-month increments proved a definite increase in returns, but the data provided no account of motivation.[171] Thus demonstrating a practical use for the case card data, whilst demanding more be collected.

Understanding failures would advance clinic services, standards and doctor-patient interactions, so a follow-up form letter to obtain the missing data was sent to patients to enquire 'whether the [contraceptive] advice [provided was] proving satisfactory'. Although the association admitted these letters may receive no response, this still provided data collection opportunities to record unanswered correspondence on patient case cards for subsequent appraisal of ongoing patient engagement. In order to implement safeguards against failures to return, North Kensington introduced 'personal visits' to patients' homes around 1930.[172] This was in effect a proactive means for the clinic to guarantee its services and prescriptions were being proficiently applied by patients, as well as to encourage the culture of compliance necessary for the contraceptive technological revolution it needed to effect its societal, medical and professionalisation goals.

Vital to the association's applied and scientific work towards standardisation was the statistical assessment of 'unaccountable failures' of method. Andrea Tone has demonstrated that by the 1930s there was a number of contraceptives known to have 'dangers and deficiencies', and so the NBCA turned its focus to assessing and determining the cause of those failures.[173] In 1936 the association requested all clinics submit figures for recorded contraceptive method failures. By October, with responses from '19 clinics and one private practice', the NBCA was confident this was a significant area for statistical study and invited clinics to analyse records and answer three questions for each 'unaccountable failure':

1. What method the patient was using or omitting to use.
2. How long they had been using it beforehand.
3. Previous fertility history with dates of pregnancies.[174]

It intended to collect and analyse details relating to 'each individual failure' to appreciate the simplicity and acceptability of contraceptive methods in actual use. This would assist doctor's prescriptive practices and, along with the regulatory pressure of the NBCA's newly minted approved list, possibly persuade some manufacturers to modify the contents or prescribed techniques of products found to be unsatisfactory.[175] It further implored Local Health Authority medical officers and private physicians to forward details of method failures experienced by patients referred to association clinics in order to maintain complete patient records and 'investigate if possible'.[176] Despite these lofty goals, Rusterholz has recently noted that while these investigations may have had value in discrete, localised studies—such as that undertaken by Wylson at the NKWWC—there was a significant communication issues that limited the unification of association-wide case card data, as the term, meaning and forms of 'failure' were understood differently across the association's various sites.[177]

This statistical work became one of two NBCA/FPA clinical research ventures spearheaded by the NKWWC undertaken between 1937 and 1960. Initial 'pilot trials' for effectiveness and acceptability assessed (original and later tailored) case cards to understand patient experiences of and with approved list contraceptives. A subsequent safety testing program began to address the long cited but never proved claim that the use of chemical contraceptives caused carcinoma of the cervix. The association oversaw clinical safety trials in conjunction with the North Kensington and Walworth clinics and unsuccessfully endeavoured to cooperate with Kings College, the Royal Free Hospital and Hammersmith Hospital in London.[178]

Late 1934, acceptability and efficiency trials began at the Walworth and North Kensington clinics. Each recruited 'patients who were willing to experiment' with a variety of contraceptives to ascertain discrete success rates.[179] Milsan chemical contraceptives were trialled first as the confident company directors promised 'an almost unlimited supply' of its products for clinical assessment. After Voge had completed his laboratory-based examinations into the contraceptive's merit and efficiency, the clinic undertook extensive statistical examinations and clinical trials to test its contraceptive ability in use, and when they were borne out by the latter, to publicise the authority of Voge's techniques and findings.[180] After Voge's departure, Baker overtook the association's laboratory testing, and from 1937 all clinical trials were for products and methods deemed effective using his testing method. This demonstrates that the association did not see its pure and applied work as distinct, but rather, quickly moved to use each to inform and extend the work of the other.

Two years later the North Kensington and Walworth clinics commenced clinical trials 'to assess the usefulness of Volpar, and other contraceptives'. Baker steered a pilot trial using case cards to track 'how many tubes or containers of each product [each clinic] distribut[ed]' and to whom. If a failure was reported, the patient's card could be consulted to determine any potentially defective spermicides.[181] The NBCA medical sub-committee approved this methodology. It is worth noting that this trial represented an applied research agenda that would ultimately have significant financial and reputational benefits for both the lead scientist, Baker, and his employers, the association and British Drug House. The project was defined and undertaken by the same people that would most profit if Volpar was proved to be the best chemical contraceptive, especially if it had been directly compared to its competitors during testing. This was certainly a win–win project for all involved.

Despite this, the product failure trials caused a dispute with Volpar's manufacturer. In 1939, the company proposed to advertise that the NKWWC clinical trials proved the product was an effective sole-use contraceptive. The NBCA refused, stressing that success in several small trials was not sufficient to guarantee this claim. In fact, the NKWWC trials contradicted the manufacturer's claimed results; of the six patients prescribed 'Volpar Gels alone', 'three ... had [shortly] reported symptoms indicating pregnancy'.[182] This obvious failure forced the association to reinforce and publicly 'reiterate its policy of recommending a double method', a barrier in conjunction with spermicide.[183] But ultimately, much to all parties' satisfaction, the association allowed British Drug House to advertise claiming that 'scientific facts backed by clinical experience show that there is every reason to believe that Volpar products are the most effective chemical contraceptive known'.[184] By 1938, many British contraceptive manufacturers co-operated with the NBCA to support the approved list as a promotional tool. Despite the association's ongoing investment in and favouritism towards Volpar, companies acceded to NBCA laboratory and clinical trials to sell merchandise, as for a short time in the 1930s, clinics, not individuals, became the primary purchasers of contraceptive products, and even when contraceptive purchase became easier and more commercial, clinics remained important sites of advertising and sales.[185]

For Gilmont products, the NKWWC's primary supplier, the emergence of Volpar was devastating. In 1941 a small-scale comparative trial that was undertaken, placed the Volpar and barrier method in direct competition with the cap and GP ointment prescription.[186] To the shock of doctors who were dedicated to the latter prescription, case card trial records dubbed Volpar superior. Wright and Malleson both reported 'patients using Volpar [with a cap had] a success rate better than [that] obtained with G.P O (100%)'.[187] The association invited the trial's expansion and its 'statisticians asked that the doctors give alternate patients Volpar and G.P.O' and record results on specially designed cards.[188] The work proceeded, however, the trial was abandoned in 1942 as many participants had to evacuate London. The Exeter clinic, where three cohorts of women were also being trialled, relayed promising findings. Some 420 women were prescribed Volpar paste or gel with a cap or a sheath for over 4893 months. From there, 46 pregnancies occurred; however, only 12 of these patients reported the contraceptive 'chemical [was] regularly used'. Direct comparison was made with 265 women prescribed other available chemicals ('G. P. ointment, Prensols, Lomolo, Milsan, Prentif Compound'). Over 4531 months 41 pregnancies occurred, 11 of which were determined 'chemical failures'. Finally, a control group of 127 women was studied, and over 10,091 marital months 375 children were conceived.[189] Whilst some doubts about the impartiality of the researchers and their leniency towards Volpar can be raised in light of these findings, there is no evidence that there was any direct bias inherent to the study.

The NKWWC effected an expanded comparative trial at two of its auxiliary clinics to determine whether the GPO method should be officially replaced with Volpar as its primary chemical prescription. Each product would be exclusively prescribed at either site and the trial records would be used to conclusively declare one more effective and acceptable.[190] These trials offered the association a model for systematically comparing contraceptive methods and products in actual use and disseminating the results as valid applied scientific research. The NKWWC and its auxiliary clinics' ability to accommodate large-scale trials enabled the association to effect several comprehensive assessments which proved more authoritative than many, smaller studies from disparate sites.

The second world war provoked severe chemical and rubber shortages which ultimately delayed the institution and resumption of clinical trials, even those based on analyses of case cards.[191] In 1943 when the association experienced great 'difficulty… getting supplies of Volpar Paste and G. P. Ointment', all tests were suspended and efforts focused on obtaining bulk supplies of still available contraceptives for distribution.[192] After the war the FPA, the NKWWC and private doctors and chemists moved to resume clinical trials as soon as possible. In relaying wartime findings, former medical sub-committee member James A. Gillison, reported that he solely prescribed Volpar Gels 'to a considerable number' of patients and determined them 'most satisfactory … [as] not one woman reported a failure … since the beginning of the war'.[193] This was an interesting and vague claim for the association to accept at face value, as it should be noted that many of the early criticisms of Marie Stopes' sociological claims were dismissed on the grounds that a failure to return or report dissatisfaction or failure did not necessarily denote either of those claims as true. However, it had to, as it had funded this 'clinical trial of Volpar' alone. In 1939 the medical committee agreed to supply the doctor with gels to distribute to 'a limited number of poor patients who did not wish to be fitted with a cap'.[194] All the results were to be personally recorded by the doctor. In 1940 the doctor reported that he had selected only 'reliable patients' for the trial and claimed that ten had already used Volpar alone for over a year.[195] Despite these apparent successes, in 1947 serious Volpar fallibilities surfaced with reports that mercury in the products damaged rubber goods. As a result, the association was forced to admit the product was known to irritate one per cent of users.[196]

In 1951 Wright claimed that North Kensington- and Exeter-based effectiveness and acceptability 'pilot trials' that compared Volpar Paste and Gel, Ortho Gynol Jelly, Elarcreme and Rendell's Pessaries, had 'proved [Volpar's] suitability' and confirmed it was the most effective spermicide on the market.[197] These trials prescribed one specific product per day, Monday to Thursday, to new patients until 50 sampled each. At the trial's conclusion, each patient completed a survey about their spermicide use to discern failures or complaints.[198] Results were compiled from case card and survey records. After the initial trial was complete, a larger trial of 500 patients was ordered.[199] Contraceptive clinical trials continued at North Kensington throughout the 1950s, and in 1960, the association requested the site spearhead a clinic-based trial of emerging oral contraceptives.[200] However, the local medical committee refused citing 'no enthusiasm regarding the present oral pill'.[201] During that decade's popular resurgence of the IUD, trials of those products were supported.[202]

A further issue the association was keen to test mid-century was that chemical contraceptives might have carcinogenic effects on the cervix. The question gained attention in 1930, prompting efforts to initiate lab-based animal trials (see below), six years later a cervical cancer smear trial began. Shortly thereafter association doctors insisted that only human trials could sufficiently investigate the matter with the urgency needed to verify spermicidal prescription was safe. Wright, then chairing the NBCA medical sub-committee, suggested a record-based study to monitor the cervixes of clinic patients over 40 who had used 'contraceptive preparations… more or less frequently, over a period of years'.[203] It employed 'Schiller's Test [to] paint the cervix with a dilute Lugols (iodine) Solution', which upon inspection 'dis[played extant epithelial cancerous cells as white patches'. Test designer Walter Schiller offered to directly evaluate any sample likely demonstrating 'early carcinomatous changes'.[204] The association defined a means of using its patient records to conduct the study and initiated the trial for patients between 35 and 44, to screen spermicide users' cervixes and a control group was used to delineate a cervical comparison.[205] In January 1938, the NKWWC and Walworth clinic each supplied the trial with 300 patients aged between 35 and 55.[206] Each participant must have used spermicides for over five years, from at least

1927. A further 300 women undergoing gynaecological surgery at free hospitals who had never used spermicides were recruited. A final group of married cervical carcinoma patients was sourced at cancer clinics. A trial case card recorded details of participant's contraceptive history, listing spermicides employed and the duration and cervical smear results[207] This program was projected to ensure the safety of NBCA-prescribed contraceptives and define a non-laboratory-based means to assess harm. It marked the association's first large-scale prospective harmlessness trial, the next studied the pill.

No data relating to the trial's outcome remains, except that the Marie Curie Cancer Centre refused to participate.[208] But in 1962 as the hormonal pill entered the British market, concerns relating to the relationship between contraceptives and carcinoma emerged again. This prompted the association to launch a larger cervical cancer smear trial. The study expanded over the decade and the FPA employed three staff doctors to perform smears as part of an association-wide harmlessness testing program.[209] This was in no small part the result of concurrent lab-based animal studies that needed to be borne out by clinic trials.

Cornering the Market: Mid-Century Contraceptive Harmlessness and Rubber Research

In 1939 the NBCA rebranded as the FPA and concurrently accepted responsibility to establish and enforce contraceptive standards through its annual approved list. The association had by then become the public face of contraception and a determined regulator of contraceptive products and practices. Around that time the NBCA/FPA expanded its efficacy research agenda by adding contraceptive safety and rubber prophylactic standards to its testing activities. It also formalised and advertised its position as an impartial regulator and explained to manufacturers exactly how it perceived its testing and approval duties.

> Any test must be arbitrary, and it does not follow that substances that fail our test are not satisfactory under natural conditions. The tests were devised to give uniformity, and in the opinion of their designer, a margin of safety.[210]

After the association's success with Baker's research and development program, it extended its research to standards testing identified by medical managers and staff as important to sufficiently guarantee approved list products and methods. To achieve this end, it must

> continue to accept this responsibility… to give the medical profession, chemists, health authorities, and the general public, guidance in judging between the various contraceptives offered for sale.[211]

Despite its enthusiasm for standardisation, the FPA repeatedly stressed that it perceived its role in the field as interim. It was ever aware that as the association's aim was to provide the public with effective contraception, it was not an ideal (or potentially even unbiased) arbiter and hoped to hand over the 'responsibility of maintaining and policing standards … [to] some authoritative and impartial body' as soon as one was willing to accept the task.[212] But until then testing responsibilities, schedules and scope increased annually and the FPA was determined to maintain its momentum and build pressure and demand for its services.

The establishment of a laboratory as the ideal site of its research and testing agenda was one current goal of the FPA; however, until it was in a financial position to rent, equip and people such a facility, it was at the mercy of external scientists and laboratories to achieve its research objectives. Nor was field testing an option, as this did not tempt the support of many individuals or groups willing to provide a site 'for conducting experiments to apprehend or control material processes'.[213] Thus the association was left to envision and commission ever 'more elaborate and thorough investigation of the biochemistry and physiology' of reproduction, which could be easily integrated into the research agenda of a sympathetic laboratory or scientist.[214] Achieving basic scientific knowledge on contraceptive safety through pointed laboratory investigations would make the field commensurate with biological and chemical studies that aimed to give approval to medical therapies and technologies for the guidance of physicians, health authorities, chemists, retailers and individual purchasers.[215] Ensuring the safety and efficacy of contraceptive products was fundamental to the FPA's function and authority, and the laboratory was a useful site for undertaking this vital work without being seen to have direct application in contraceptive clinics, even if it ultimately did.

General harmlessness testing originated with the Birth Control Investigation Committee. It quickly recognised a potential risk from applying chemical contraceptives directly into the vagina or being used in conjunction with a diaphragm/cap. It was theorised that in protracted use, these might cause carcinoma or sterility. Contraceptive safety was also on Cecil Voge's original 1930s testing agenda to allay physician's fears about the 'medical harmfulness of contraceptive practices'.[216] As the association provided these products to consumers daily, it needed to examine these claims and to disavow products that were determined harmful.[217] Ensuring the safety of chemical contraceptives was heavily funded from the late 1930s.[218]

In its 1930 Draft Statement, the BCIC defined its ideal approach to safety testing. Animals would be the initial primary subjects as 'work of this nature can be done on a large scale… and then the more important points checked on human[s]'. This stance was carried into later NBCA/FPA harmlessness research.[219] Baker's chemical studies instigated the council's concern with safety and prompted a collaboration with histologist Harry Carleton and pharmacologist and pathologist Howard Florey to explore the use of 'experimental animals for testing the harmful effects of spermicidal compounds'.[220] This group determined bitches to be the most appropriate trial animal, and in 1930 a base procedure for evaluating contraceptive safety was defined by Baker, Carleton, anatomist Solly Zuckerman and research assistant Clare Harvey. The premise was that three medium sized bitches would be injected daily with oestrone and ethyl oleate until their 'vaginal smears showed no pus cells'. Then a chemical contraceptive would be inserted into the vagina every day for a fortnight along with daily oestrone injections. The day after the final application, the bitches would be killed, and a full histology of their reproductive systems executed.[221] Later, when formal harmlessness experiments were being designed, the association agreed that 'substances which appear to have no harmful effect on animals can then safely be tried out on a small scale on women in clinics'.[222]

Harmlessness testing became an NBCA medical sub-committee imperative after Carleton stated in introducing *The Chemical Control of Contraception* (1935), that

> the inclusion of a paste or jelly between the os uteri and a cervical cap is to be regarded as a potentially dangerous procedure, and that women adopting this practice should not so much be encouraged by experts as warned of its eventual pathological possibilities.[223]

Upon realising that no formal consideration had been paid to the possible 'harmful results' of repeated application of chemicals into the cervix over protracted periods,[224] the committee approached learned doctors regarding this possible danger. In January 1936 Carleton, Florey and cancer researcher and pathologist Beatrice Pullinger were invited to discuss the issue. The lack of scientific awareness of the effects of protracted, repeated cervical application of chemicals stemmed from two problems: the 'time lag necessary in the production of a carcinoma [and] the sporadic nature of the carcinogenic agents'. Pullinger offered to furnish the committee with a list of 'established carcinogenic agents' but stressed this was not exhaustive as new toxic compounds were constantly being discovered. She was essentially flagging the need to design and undertake a novel protracted and prospective study to consider the existence of a potentially toxic relationship. As this had the potential to significantly impact public health, Pullinger and the association (that had no intention of halting its contraceptive work during the investigation) acknowledged that it was important to implement a practical research agenda. So two options were suggested: the employment of a statistician to investigate the matter through clinic patient data and/or a lab-based experimental program.[225] The latter offered the association the greatest opportunity to undertake pure chemical and biological safety research, and thus to formally integrate harmlessness testing into the realm of contraceptive science. The NBCA medical committee agreed 'it would be valuable to obtain scientific evidence to support this ... by examining women who have practised birth control for several years and who have now reached or passed through the age at which trouble is most likely to arise'.[226] It was suggested that the 'most satisfactory way of excluding this danger' would be to implement a test applying contraceptive chemicals to animal cervixes for gradated time periods.[227] Despite approving these ideals of safety testing, harmlessness never received wholehearted association support or funding until the conclusion of the second world war as its limited resources were habitually deferred to its spermicidal research and testing. Harvey, the only researcher the NBCA retained from its early theoretical safety work, transitioned to its spermicidal program in the late 1930s when Baker left to work for British Drug House.[228] But the FPA was 'kept busy... war or no war', and when the workload accumulated owing to wartime pressures, its regulatory controls proved to be increasingly pliable.[229]

A complete overhaul of the FPA's standardised testing programs was instituted post-war. Its medical sub-committee identified flaws in its early testing methods and attempted to improve them and institute new ones. The association maintained the centrality of innovative scientific standards to its regulatory work and was conscious to periodically assess testing methods to ensure ongoing efficacy and implement improvements when necessary. The Baker Test of spermicidal efficiency remained acceptable.[230] However, Baker's Diffusion Test, which assessed spermicides rate of dispersal at body temperature, was deemed insufficient as it was difficult to reproduce, scientifically debateable, and caused considerable unrest between the FPA and manufacturers[231]; a not uncommon issue, but an added hassle for the association in negotiating standards and commercial arrangements.[232] Failing this test was a frequent reason chemical contraceptives were rejected from the approved list or relegated to sections B and C, where products that had yet to pass all Baker's tests or be proven in clinic were itemised.[233] Despite his repeated assertion that diffusion tests were 'unnecessary, and… the major approved list "snag"', seminologist and sub-fertility expert Hans A. Davidson was employed to develop a new and more appropriate '"contact" test and standard'.[234] The Davidson Contact Test in which equal volumes of sperm and contraceptives were held in an entirely sealed slide until the former were all immobilised, was accepted as the association's standard from August 1953.[235] After that no further developmental or revision work was undertaken on spermicides by the FPA, although its dedication to seminological testing remained high.

From 1948, the FPA pushed for 'a series of harmlessness tests on all the chemical contraceptives on the approved list'. The association initially approached Harvey to develop a method of testing safety. She rejected Baker's proposed approach as 'a waste of a good bitch' and clarified 'I wouldn't have taken on [safety testing] for all the gold of the Incas'.[236] Later she was persuaded to forward specifications of how she would hypothetically 'tackle it'. Harvey advised an applied scientific investigation into chemical contraceptive acceptability using FPA clinic volunteers. These women could be separated into groups and asked to test a product and report any irritation. If every other probable source of irritation could be eliminated, it could be safely concluded the contraceptive was the cause. Harvey stressed that for many women irritation and messiness were interchangeable and so the subjective nature of contraceptive preference would have to be considered in designing the trials.[237] Respected physician and analyst Carlos Blacker also supported a clinic-based statistical analysis of 50 human volunteers having 'passed the menopause ... who would agree to insert a quinine pessary every night for a month'. He concurred this method would provide a 'strong indication', if not actual proof, of the general acceptability and harmlessness of tested products.[238] The medical sub-committee, however, disagreed. They found that conducting animal rather than human trials presented the least risk and promoted laboratory-based testing over clinic-based methods.

Finding a suitable researcher and methodology for its safety goals was an arduous process, and there were suggestions that 'tests [might have] to be abandoned' until a casual discussion with Zuckerman uncovered a feasible candidate. Professor of Reproductive Endocrinology Peter Eckstein, worked in the Anatomy Department at the University of Birmingham. Eckstein was 'willing and able to undertake such tests using rhesus monkeys' and promised to develop a test to assess spermicidal safety.[239] In 1953, Eckstein offered to trial his proposed method even before receiving any funds. He guaranteed medical staff that the project could be achieved 'relatively easily [with]in the framework of existing research'.[240] After six months study, Eckstein 'tentatively' described that none of the three products being tested (duro-creme, ortho-creme and ortho-gynol) 'had a marked deleterious effect on the vagina'. But he asked for a delay in advertising these findings until his method was perfected.[241] The medical sub-committee was considerate of the slow pace inevitably involved when working within the constraints of the menstrual cycle and remained satisfied as long as 'work was progressing' and Volpar received appropriate attention.[242]

Late 1954, Eckstein's scheme to test each product's safety on three monkeys was formalised. However, a serious study that could directly inform association practice and prescription could not yet be effected as Eckstein discovered that not enough was known about the normal or baseline conditions of monkey vaginas to reliably discern safety.[243] As a result the scope and cost of harmlessness research expanded significantly and the cash-strapped FPA had to apply for Eugenics Society funds to admit the essential preliminary research in order to avoid diverting funds from direct harmlessness testing.[244]

Eckstein proposed to undertake control tests on three to four monkeys, performing vaginal histology and biopsies of each over several months of their normal menstrual function to establish a solid baseline for testing and assessing safety.[245] Following preliminary testing, one of the sixteen purchased monkeys would have the arbitrary amount of a given chemical inserted directly into the vagina once daily for a set period. During that time biopsies would be taken: one prior to commencement, two at different phases of application, and one at conclusion.[246] Eckstein proposed that biopsies collected from three or four animals would constitute a control test.[247] The Eugenics Society granted £300 over three years to purchase and house monkeys and to fund three rounds of testing over nine months with regular 'progress reports'.[248] This pragmatic and cooperative approach was another example of the shared ideological goals of these groups and their unrelenting belief in the value of pure scientific knowledge to inform applied scientific ends.

Eckstein's first official report was presented late 1954. It related the results of the first three products tested and concluded that three out of four monkeys tested stayed entirely healthy and the fourth recovered following treatment and showed no ongoing effect. Vaginal inspections revealed that epitheliums thinned, cornified slightly, and were easily traumatised by invasive examination, but biopsies revealed 'no evidence of any harmful effect'.[249] The subsequent 1955 report was more extensive and authoritative as it succeeded the base tests. Seven contraceptives were examined including the entire Volpar range. The results were promising, and all monkeys tolerated the contraceptives well, some daily for up to seven months. However, all monkeys experienced diminished haemoglobin and red blood cell counts and experienced longer, heavier and more irregular periods. Eckstein's earlier safety results were sustained, and he reported that eight of the nine animals exhibited 'no significant lesions on vaginal biopsy'. At the experiment's conclusion, two animals were killed, and their entire reproductive tract excised and inspected. 'No pathological condition of any part of the necessary reproductive tract' was exposed.[250] Eckstein's test was judged successful and applied to all products manufacturers submitted for approved list inclusion after 1956. Any product, not safety tested was relegated to Section B.[251] An effective scientific method was thus defined to assess the chemical, biological and medical effects of contraceptives and an acceptable standard of safety defined and enforced. Not long after this in 1957, the International Planned Parenthood Federation also spearheaded and assessed cervical smear test studies as part of its international investigations into harmlessness. The 'Cap Test' that placed a spermicide slathered cap in place for 24 hours and assessed patient's smear results at the beginning and end was trialled by Dr. Melba Kamat at the Contraceptive Testing Unit at the

Indian Government Testing Centre in Bombay[252]; and a further study based on repeated application of its spermicides and analysis of smears was developed by Dr. Carl Hartman in the USA for Ortho Products.[253] Thus, in the post-war world the UK was no longer the sole epicentre of contraceptive standards work and the FPA was able to export its research methods to and contrast and cooperate with sympathetic groups and advocates further afield in the hope of effecting greater acceptance and reliance on the contraceptive scientific work it had pioneered in the not-for-profit space.

Concurrent with the foundation of safety testing in the mid-1930s, another means of legitimising contraceptive science and products was established: the chemical appraisal and standardisation of rubber prophylactic quality and manufacture.[254] The NBCA medical committee engaged rubber experts across Britain to design and spearhead its rubber standards program.[255] The agenda began under fermentation chemist and rubber refinement pioneer Philip Schidrowitz. In 1935 he produced *Tentative Standards Specification A*, the first rubber contraceptive standard that the NBCA disseminated to manufacturers as a quality and efficacy guideline. These standards were updated and improved mid-1937.[256]

Amy Slaton has argued that the establishment of manufacturing and product standards was a feature of the industrialised world since the turn of the twentieth century. From then, Slaton claims, standards and specifications were increasingly common, constituting a 'technical and legal communication' from scientific experts to their clients in manufacturing and government assuring quality. This benefitted scientists by creating new domains necessitating scientific influence and authority, and manufacturers through the establishment of public trust in their products and brand. Standards articulated the optimal quality, grade and/or size of a product and specifications often incorporated standards, to define the parameters of a product or a method of practice. Slaton further contends that manufacturers increasingly embraced the potential benefits of intercompany regulation, whereby technical knowledge was shared, and investigative labour was used economically without duplication. This was primarily achieved through the foundation of 'private, proactive, centralised' bodies, willing to define and police standards and specifications for industrial products and practice.[257] Rubber contraceptive products were no exception to these engineering and manufacturing conditions, and their standardisation and regulation in Britain was a task the NBCA/FPA was eager to implement and guarantee.

From 1935 all the major British rubber contraceptive manufacturers—London Rubber Company, Lamberts Prorace Ltd., later Lamberts (Dalston) Ltd., Burge, Warren and Ridgely, Ortho Pharmaceutical Ltd., and finally, Prentif Ltd.—allied with the association and worked to meet the standards it laid down regarding washable sheaths, condoms/thin sheaths, diaphragms and cervical caps. Initially these standards specified that all products must pass inflation tests, though no specifics were articulated regarding how far they were to be inflated or for how long; elongation tests to 900 per cent length prior to break[258]; a test of tensile strength; and finally a visual assessment to ensure the product displayed no flaws or splits.[259] Then an artificial ageing process was achieved by boiling products at 70°C for between one and three days depending on rubber thickness and all tests were repeated to confirm products didn't degrade over time.[260] Schidrowitz's tentative guidelines insisted dating was imperative for all approved list rubber products, and so the association insisted manufacturers stamp packaging with a date of manufacture and/or expiry. Many early tests only approved products 'subject to dating'.[261]

From September 1935 the association, through Schidrowitz, instituted a large-scale rubber testing program. Its schedule met little objection except from London Rubber, who in February 1937 sent a thorough criticism of the association's testing rigour and methods. The manufacturer withdrew its concerns upon learning its products 'stood the test quite well'.[262]

Manufacturers found the association's insistence on dating its most contentious regulatory demand. But even that seemed solved by late 1937 when the manufacturers of all approved list products agreed to adhere to a maximum three-year expiration date, with the exception of some products whose lifespan was proved shorter by Schidrowitz's tests. Following a meeting rejecting the NBCA medical committee's decision 'that the 3 years' guarantee could not be lengthened', all manufacturers arranged to modify packaging to reflect this resolve. The association, demonstrating its dedication to the maintenance of good working relationships as well as high standards, sent a notice to retailers regarding the decreased rubber life span from five to three years, as it worries that failing to do so might prove 'damaging to [a manufacturer's] reputation' and impact trust in the practice.[263]

In November 1937, the Director of Research for the Rubber Research Association, B. D. Porritt, met with the NBCA testing committee to criticise the association's regulatory approach and timing. Porritt refused to endorse Schidrowitz's testing methods, particularly those for ageing, and insisted the association would have great difficulty 'setting standards for rubber contraceptives as the whole trade was in a state of transition from old methods to new' owing to the development and use of latex.[264] But Porritt conceded that 'Schidrowitz was one of the only two experts in this matter' and the Rubber Research Association had never 'tested such goods'.[265] He promised that the Rubber Research Association was 'willing to undertake' the necessary 'special research [needed] to evolve' a truly sufficient methodology to meet the association's needs. The NBCA was eager to trust the promises of what it deemed to be a potential ally and quickly forwarded rubber samples to Porritt for him to devise a test and send results that could be compared with Schidrowitz's.[266] The NBCA's optimism that a semi-official research organisation was willing to undertake contraceptive research and standardisation was short lived, as Porritt's response betrayed his real aim which was to solicit a six-month commission of £150 from the association to undertake 'thorough research into the question of standards for rubber contraceptives'. Porritt had not explained this at their meeting and the allocation of such a vast amount was impossible for the young not-for-profit organisation, especially as Porritt had given it no cause to doubt its current methodology.[267]

The war necessitated the relaxation of contraceptive standards. During those years rubber goods were assessed understanding that 'manufacturers have had to resort to [a lesser] type of rubber owing to the war conditions'.[268] Rubber test results from 1942 to 1948 were published with the disclaimer: 'having regard to the present circumstances may be regarded as reasonably satisfactory'.[269] In 1940, Lamberts Dutch Caps were twice submitted for testing and failed badly both times. The association notified Lamberts its present manufacturing standards were inadequate. Lambert's conceded this fact but could not entertain making any changes until the war's conclusion. The association resolved that as Lambert's caps had proved satisfactory in clinic use and were one of the few caps still presently available, it would 'leave them on the list'.[270] Borge has shown that this approach was actually not uncommon for the association, as it had to balance patient's contraceptive needs with its professional and regulatory relationships with manufacturers, most of whom were also deeply invested in legitimising the NBCA/FPA as a means of broadening their own influence and sales figures.[271] However, this common need and benefit at no point obliged a subordination of any parties' business interests for the wider movement.

In 1941 London Rubber ceased dating. It worried that if consumers knew products were made with the inferior wartime rubber, that merchandise would not be saleable. This was a legitimate concern, but arguably one that could be negotiated with the association, as had been done for Lamberts. However, the relationship between London Rubber and the FPA was significantly different. As Borge and Jones have argued, it was more contentious and far less grace and trust had been established between the parties, than had been achieved with other manufacturers. As a result, the association responded to the removal of dating far more zealously, as this was directly contrary to the negotiated arrangement whereby it became the prime consumer and guarantor of rubber goods and would publicly distribute them if its standards of manufacture and quality were met.[272] The FPA responded by removing undated London Rubber products from its newly printed approved list and admonished the company by 'put[ting] pen through the [London Rubber] Co. Caps *and* their address' on the already 'rolled off… copies'.[273] This relationship underwent significant changes over the subsequent two decades and ultimately London Rubber earned the association's trust and loyalty. It tailored its cap and diaphragm production specifically to suit the association's female-centred prescriptions, funded some of the association's promotional activities and essentially built a dependence on its rubber goods; ultimately at the expense of the other manufacturers within the FPA orbit. As London Rubber increased its share of the local contraceptive market, the power dynamics shifted and later disputes over association regulatory rules were ultimately resolved in the manufacturers favour to maintain access to below cost rubber contraceptives in the era of over the counter and even vending machine condom sales.[274]

The FPA undertook annual rubber testing throughout the 1940s, but owing to rubber shortages and quality concerns, the practice was not formalised until 1954.[275] From 1950 it sought to impose increasingly rigid testing and quality standards as it saw the scientific community was getting ready to accept the necessity of rubber standards. It was accepted the FPA was then undertaking 'successful work in this field' and its method was worth studying and possibly applying to other products and industrial organisations.[276] Importantly, in 1953 some commercial demand for FPA approval came from chemist shops following the assertion of the statutory regulator of poisons and pharmacists, the Pharmaceutical Society of Great Britain, that chemists falling under its control were limited in how they advertised contraceptives in store by displaying 'an authorised notice that articles approved by the Family Planning Association or other body approved by the Council are sold'.[277] In light of these gains, the association admitted in principle that its methods were not yet perfect but acknowledged it was 'still in the process of answering questions' of method and scope and was still hoping to pass on the work to an official body when possible. The association's regulatory capacity did not encompass the entire contraceptive industry, rather it was always limited to those manufacturers who desired FPA affiliation. This revelation demonstrates the symbiotic nature of the contraceptive regulatory process the FPA had established. It was not at any point a formalised, authoritative and independent arbitrator of an entire industry, but rather sat somewhat uncomfortably in a self-defined niche somewhere between primary client and product promoter. The limited authority the association was able to wield in this relationship owed to it being the discriminating and trusted middle-man in the sales and standards chain-of-command between manufacturer and user. This reality explains the association's consistent enthusiasm at the prospect of passing its methodology to some government-affiliated authority—that could compel the regulation that under the FPA's authority, was a matter of agreement and cooperation.[278]

Though the association was disappointed with the narrow scope of its regulatory sway, its dedication to the principle of standardising contraceptives only increased. In 1953, the FPA instituted higher standards for rubber goods via Schidrowitz's revised 'Specification of Tests &C. of Rubber Goods'. Condom and sheath inflation tests were formalised with a 'minimum diameter at maximum point of inflation' and a standardised test time frame of fifteen minutes. The quality of the rubber pre/post ageing had to hold 1500/1200 lbs., and its elongation at break must exceed 800/720%. Ageing now demanded submersion in boiling water for thirty hours. All caps must stretch over a smooth mandrel to at least three times its size without tearing and a seventy-two-hour ageing test was required to guarantee long-term useability. All rubber products now required both a date of manufacture and discard.[279]

Rubber goods' sizes were officially measured and standardised through the association's efforts and specifications. Caps and diaphragms were the prime products of the association; they were political and feminist in nature, but as other historians have demonstrated, an absolute money hole for all parties.[280] They were arduous to produce and test, easily fallible in application and use and had to be made in a large variety of shapes and sizes meaning there was always a risk of over or under production and stock. Many companies worked with the FPA to produce appropriate caps, but ultimately the London Rubber Company also dominated this market owing to its 'enviable ability to produce [them] quickly and at great economies of scale'.[281] Caps and diaphragms demanded specific focus and debate regarding the ideal spring tension of the rim as in actual use many had proved too tight and snapped upon particularly vigorous folding, insertion or removal. The FPA was happy with the spring tension in small and medium sized caps,[282] but worried more was needed for its larger varieties.[283] Sizing standards were also a central concern regarding the three condoms and washable sheath sizes. Mid-1950s sizes were scaled down significantly to reflect reality: small (<6½ inch), medium (7 inch) and large (>7½ inch). All clinics were advised medium was the standard and other sizes must only be dispensed in 'exceptional circumstances'.[284] Unlike for caps there was no opportunity for fitting these, and so it was safer to err on the side of caution rather than accept a customer's guesstimate as gospel.

It's worth noting that although the FPA prescribed female rubber methods, either the cap or diaphragm, to almost every patient that attended the clinic into the 1960s, much of its rubber regulation schedule was dedicated to assessing and approving condoms and sheaths. This was in part because men were initially more likely to approve of and purchase contraceptives, but also because the volume of condoms sold, either in clinics or far more commonly in chemist shops and via catalogue or vending machine, over this period was significantly higher.[285] Condoms and sheaths needed to be replaced far more frequently than caps, and so guaranteeing them, even if they were not the association's preferred method was vital to the overall trustworthiness of the approved list.

In the mid-1950s the FPA also claimed authority concerning pricing and cleaning instructions for washable rubber contraceptives. The association, already disillusioned by the quality of Lamberts (Dalston) products, strongly disagreed with the phrasing of its cleaning instructions for Lambutt Caps. The manufacturer advised that 'rubbing' would suffice for washing; but the FPA medical committee recognised this 'would undoubtedly be too gently interpreted', and as 'mucus and rubber [were] not easily separated' insisted the wording be altered to say that 'after use, this cap should be scrubbed with a nail brush and a neutral soap'.[286] Lamberts fearing 'this cleansing method [could] be abused', but also worried that the association might drop its products, agreed to a trial revision, that it would rescind if aggressive cleaning impacted the integrity of its products.[287]

Lamberts came under further fire when it was discovered that it had met the FPA's demand that 'cheapness was the first consideration', with reduced quality products when compared with those sold at Lambert's recommended retail price in regular chemists.[288] The association was incensed its clients had to 'renew their caps every six months' and claimed this was false economy, especially for poorer users. Lamberts conceded some caps sold at 'special clinic rates' had been returned, but alleged clinic staff were to blame for 'not carrying out the correct procedures in testing [and handling] the caps'; its proof was that clinics had returned caps in a 'very mutilated condition' with 'several marks in the rubber caused by fingernails'.[289] Even with the best protocols and instructions, the association had no way to guarantee perfect conduct and use in clinics or homes and in the interest of maintaining good relations, the FPA withdrew its charge. The reality was that by this point most manufacturers were being squeezed out of the market by the dominant London Rubber Company. Its goods were made at a consistently high quality and were priced so cheaply, that other companies were unable to compete without cutting corners on quality or selling at a loss.[290] It was the end of the competitive age of rubber, standards had been set, political and financial deals done, and publicising its achievements and services was the FPA's end game.

Conclusion

Between 1930 and the mid-1950s the NBCA/FPA implemented a series of scientific projects and programs that were intended to inform and guarantee its own practices and products and standardise the broader contraceptive industry. As has been discussed, this agenda was broad and incorporated both pure and applied science using laboratory and clinic-based studies and data to make decisions and recommendations about manufacturers, methods, and products, and their results in simulated and actual practice. The research agenda was often fraught with political and ethical challenges that needed to blend the association and manufacturer's mutual desire for legitimacy with their contrasting financial needs and motivations, and the safety, efficacy and acceptability needs of end users.

The research agenda made some important gains during the 1930s with the development of scientific methods to assess and standardise the spermicidal abilities of specific chemical products, the establishment and annual revision of the approved list of contraceptives and analysis of clinic data to understand contraceptive acceptability for users, failures of products and for follow up. However, the expansion of the sociological and lab-based studies really took off in the post-war period when numbers of FPA clinics expanded at a pace and impetus, experts, and funds were available to pursue this type of work.

Between 1940 and 1956, increasingly stringent testing and reporting directly influenced which products were included on the annual approved list and where. With the institution of reliable and reproducible harmlessness and rubber tests to complement its efficacy tests, the association was able to set and enforce minimum standards for products it supplied. Further, as the association was both the regulatory authority and a prime contraceptive consumer and disseminator, many manufacturers were willing to pay ever-increasing testing fees to remain an approved manufacturer and have their products prescribed in FPA clinics. Ultimately, the introduction of mandatory harmlessness testing streamlined the list and imbued each featured contraceptive uniform status as safe, effective and reasonably priced.[291] After 1950, the FPA improved and extended its testing to induce British regulators to oversee rubber and chemical contraceptives. Allying its testing methods with those of the British Standards Institution which assessed rubber goods, and the Pharmacopoeia Commission that judged chemicals was the FPA's final regulatory goal.

Notes

1. NBCA Medical Sub-Committee Minutes, Session One, 17 November 1934. Wellcome Library, Archives of the Family Planning Association (WL/SA/FPA), WL/SA/FPA/A5/88.
2. Report on FPA List of Approved Contraceptives, 1953. WL/SA/FPA/A7/16.
3. *FPA Approved List*, June 1952. WL/SA/FPA/A7/3. This is the first claim that the FPA approved list filled a perceived gap regarding contraceptive product standards. It asserts 'no contraceptive products are listed

in the British Pharmacopoeia but for many years the Family Planning Association has investigated the qualities of proprietary contraceptives and made the findings available to interested enquirers'.

4. *NBCA Approved List*, October 1937. WL/SA/FPA/A7/5. A prior draft list dated October 1936 exists, but this copy is littered with amendments and I do not consider it the first official list.
5. John Loadman, *Tears of the Tree: The Story of Rubber—A Modern Marvel* (Oxford: Oxford University Press, 2005), 35.
6. Ann Collier, *The Humble Little Condom: A History* (New York: Prometheus Books, 2007), 134; Adele E. Clarke, *Disciplining Reproduction: Modernity, American Life Sciences, and the 'Problems of Sex'* (Los Angeles: University of California Press, 1998), 167.
7. Clarke, *Disciplining Reproduction*, 167.
8. Loadman, *Tears of the Tree*, 293–311.
9. George Standring and William Reynolds, *The Malthusian Handbook: Designed to Induce Married People to Limit Their Families within Their Means*, Fourth Edition (London: W. H. Reynolds, 1898), 46.
10. *NBCA Approved List*, October 1937. WL/SA/FPA/A7/5.
11. *FPA Approved List*, February 1950. WL/SA/FPA/A7/5.
12. *FPA Approved List*, February 1950. WL/SA/FPA/A7/5.
13. Amy Slaton, '"As near as practicable": Precision, Ambiguity, and the Social Features of Industrial Quality Control', *Technology and Culture* 42, 1 (2001): 51–52.
14. NBCA Prices of Contraceptives, 22 February 1939. WL/SA/FPA/A7/5.
15. Humphry Rolleston, 'Birth Control Investigation Committee', *British Medical Journal* 2 (1927): 805–6.
16. T. D. Whittet, 'Drug Control in Britain: From World War 1 to the Medicine Bill of 1968', in *Safeguarding the Public: Historical Aspects of Drug Control*, ed. John B. Blake (Baltimore: The Johns Hopkins Press, 1970), 31–2.
17. Whittet, 'Drug Control in Britain', 27–30.
18. 'Rolleston, 'Birth Control Investigation Committee', 805–6.
19. Arthur E. Giles, 'The Need for Medical Teaching on Birth Control', *Lancet* 1, 5395 (1927): 165–7; 'National Birth Control Association', *Lancet* 2, 5809 (1934): 1456.
20. Qualified approval is given 'only after one, two or three children', 'only on medical advice', and 'only on health grounds'. John Peel, 'Contraception and the Medical Profession', *Population Studies* 18, 2 (1964): 138–9.
21. Peel, 'Contraception and the Medical Profession', 138–9.
22. Caroline Rusterholz, *Women's Medicine: Sex, Family Planning and British Female Doctors in Transnational Perspective, 1920–70* (Manchester, Manchester University Press, 2020), 41–2.

23. *The Practitioner: Special Number on Contraception*, July 1923.
24. Peel, 'Contraception and the Medical Profession', 140.
25. 'Birth Control', *Lancet* 2, 5577 (1930): 147.
26. 'Problems of Birth Control', *Lancet* 2, 5586 (1930): 650.
27. Arthur Ellis, et al., 'A New Chemical Contraceptive', *Lancet* 2, 6008 (1938): 970.
28. Ilana Löwy, '"Sexual Chemistry" before the Pill: Science, Industry and Chemical Contraceptives, 1920–1960', *British Journal for the History of Science* 44, 2 (2011): 245–74.
29. Merriley Borell, 'Biologists and the Promotion of Birth Control Research, 1918–1938', *Journal of the History of Biology* 20, 1 (1987): 64.
30. Clarke, *Disciplining Reproduction*, 3.
31. Memorandum 'Sex Committee—National Research Council—Hamilton Project, 12 March 1928. Rockefeller Archive Centre, Bureau of Social Hygiene Papers (RA/BSHP). Quoted in Borell, 'Birth Control Research', 69.
32. Borell, 'Birth Control Research', 75.
33. Clarke, *Disciplining Reproduction*, 200–4.
34. Jessica Borge, *Protective Practices: The London Rubber Company and the Condom Business* (Montreal & Kingston: McGill-Queen's University Press, 2020); Claire L. Jones, *The Business of Birth Control* (Manchester: Manchester University Press, 2020); Natasha Szuhan, 'Sex in the Laboratory: The Family Planning Association and Contraceptive Science in Britain, 1929–1959', *British Journal for the History of Science* 51, 3 (2018): 487–510.
35. Letter I. James to A. Stone, 10 February 1953. WL/SA/FPA/A7/16; Response A. Stone to I. James, 19 February 1953. WL/SA/FPA/A7/16.
36. Cecil Voge, *The Chemistry and Physics of Contraceptives* (London: J. Cape Publishing, 1933), 224; Borell, 'Birth Control Research', 73.
37. Löwy, '"Sexual Chemistry"', 250–3.
38. Robert Latou Dickinson and Louise Stevens Bryant, *Control of Conception: An Illustrated Medical Manual* (Baltimore: The Williams and Wilkins Company, 1931), 37.
39. John Reed, *The Birth Control Movement and American Society: From Private Vice to Public Virtue* (Princeton: Princeton University Press, 1978), 242.
40. Voge, *The Chemistry and Physics of Contraceptives*, 22–5.
41. Voge, *The Chemistry and Physics of Contraceptives*, 53.
42. Borell, 'Birth Control Research', 72.
43. Dickinson and Bryant, *Control of Conception*, 37. This is a verbatim transcript of the goals Voge described upon undertaking his research.

44. Voge, *The Chemistry and Physics of Contraceptives*, 22.
45. Voge, *The Chemistry and Physics of Contraceptives*, 23–4.
46. Voge, *The Chemistry and Physics of Contraceptives*, 224.
47. Voge, *The Chemistry and Physics of Contraceptives*, 225.
48. Voge, *The Chemistry and Physics of Contraceptives*, 24.
49. North Kensington Medical Committee Report, 15 November 1934. WL/SA/FPA/NK225.
50. John R. Baker, 'The Spermicidal Powers of Chemical Contraceptives: II. Pure Substances', *Journal of Hygiene* 31, 2 (1931): 211.
51. BCIC Draft Statement of Work,1930. WL/SA/FPA/A13/5.
52. BCIC Summary of Activities, 1930. WL/SA/FPA/A13/5; BCIC Present Scope of Work; 1930. WL/SA/FPA/A13/5.
53. John R. Baker, 'The Spermicidal Powers of Chemical Contraceptives: VI. An Improved Test for Suppositories', *Journal of Hygiene* 34, 4 (1934): 474; Baker, 'II. Pure Substances', 189–90.
54. John R. Baker, R. M. Ranson and J. Tynen, 'The Spermicidal Powers of Chemical Contraceptives: VII. Approved Tests', *Journal of Hygiene* 37, 3 (1937): 474.
55. Baker, 'II. Pure Substances', 192–3.
56. Baker, 'II. Pure Substances', 195–6.
57. John R. Baker, 'The Spermicidal Powers of Chemical Contraceptives: I. Introduction, and Experiments on Guinea-Pig Sperms', *Journal of Hygiene* 29, 3 (1929): 326–8; John R. Baker, 'The Spermicidal Powers of Chemical Contraceptives: III. Pessaries', *Journal of Hygiene*, 31, 3 (1931): 313–6; Baker, 'II. Pure Substances', 192–7.
58. John R. Baker, 'The Spermicidal Powers of Chemical Contraceptives: IV. More Pure Substances', *Journal of Hygiene* 32, 2 (1932): 172.
59. Joan Malleson, *The Principles of Contraception: A Handbook for General Practitioners* (London: Victor Gollancz Ltd., 1935), 156; Gladys M. Cox, *Clinical Contraception*, Second Edition (London: William Heinemann Ltd., 1937), 134.
60. Voge, *The Chemistry and Physics of Contraceptives*, 164.
61. NBCA Medical Sub-Committee Minutes, Second Session, 4 May 1935. Wellcome Library, Archives of the Eugenics Society (WL/SA/EUG), WL/SA/EUG/D/12/16/22.
62. NKWWC Medical Committee Minutes, 26 September 1935. WL/SA/FPA/NK225.
63. Richard Soloway, 'The "Perfect Contraceptive": Eugenics and Birth Control Research in Britain and America in the Interwar Years', *Journal of Contemporary History* 30, 4 (1995): 652.
64. Borell, 'Birth Control Research', 73.
65. Adele E. Clarke, 'Maverick Scientists and the Production of Contraceptives, 1915–2000+', in *Bodies of Technology: Women's Involvement with*

Reproductive Medicine, eds. Ann Rudinow Saetnan, Nelly Oudshoorn and Marta Kirejczyk (Columbus: Ohio State University Press, 2000), 45; Soloway, 'The "Perfect Contraceptive", 652.
66. Baker, 'I. Experiments on Guinea-Pig Sperms', 323; Baker, 'II. Pure Substances', 192; Clarke, *Disciplining Reproduction*, 188.
67. Letter R. Squier to C. P. Blacker, 5 January 1938. WL/SA/EUG/D/12/26; Letter L. S. Bryant to R. L. Dickinson, 18 March 1933. Harvard Library: Countway, National Committee on Maternal Health Papers, Box 17, Folder 575; Soloway, 'The "Perfect Contraceptive"', 641, 652.
68. Clarke, *Disciplining Reproduction*, 186–90.
69. *The Contact Test: A Method of Measuring Spermicidal Actions*, 14 April 1953. WL/SA/FPA/A7/15.
70. British Pharmacopoeia Commission Ad Hoc Committee on Spermicides Minutes, 15 September 1958. WL/SA/FPA/A7/22.
71. Löwy, 'Sexual Chemistry', 259.
72. Baker, 'II. Pure Substances', 189.
73. J. R. Baker and M. V. Bowler, Interview Minutes, 5 February 1935. WL/SA/FPA/A7/13/1.
74. Letter C. Harvey to L. Arnold, 1 November 1950. WL/SA/FPA/A7/16.
75. J. R. Baker and M. V. Bowler, Interview Minutes, 5 February 1935. WL/SA/FPA/A7/13/1.
76. *National Birth Control Association Approved List*. [c.1949]. WL/SA/FPA/A7/5.
77. Letter Messers. Coates and Cooper Ltd to I. James, 12 July 1951. WL/SA/FPA/A7/13.1.
78. Letter Gilmont Products Limited to J. R. Baker, 19 February 1935. WL/SA/FPA/A7/13.1.
79. Letter Gilmont Products Limited to H. Holland, 1 March 1935. WL/SA/FPA/A7/13.1.
80. Letter Gilmont Products Limited to H. Holland, 1 March 1935. WL/SA/FPA/A7/13.1.
81. Clarke, *Disciplining Reproduction*, 188.
82. Letter NBCA to Gilmont Products 19 October 1938. WL/SA/FPA/A7/13.1.
83. Response Gilmont Products to NBCA, 27 October 1938. WL/SA/FPA/A7/13.1.
84. Szuhan, 'Sex in the Laboratory', 502.
85. Gilmont Products Chief Chemist Report on G. P. Ointment, January 1949. WL/SA/FPA/A7/13.1.
86. Response Gilmont Products to NBCA, 27 October 1938. WL/SA/FPA/A7/13.1.

87. J. R. Baker and M. V. Bowler, Interview Minutes, 5 February 1935. WL/SA/FPA/A7/13.1.
88. Dickinson and Bryant, *Control of Conception*, 36–7.
89. Response Gilmont Products to NBCA, 27 October 1938. WL/SA/FPA/A7/13.1.
90. Szuhan, 'Sex in the Laboratory', 502.
91. Graeme Gooday, 'Liars, Experts and Authorities', *History of Science* 46, 154 (2008): 431–56.
92. Response H. Holland to Gilmont Products 6 February 1942. WL/SA/FPA/A7/13.1.
93. H. Holland and Gilmont Products Meeting Report, 12 March 1942. WL/SA/FPA/A7/13.1.
94. *Approved List*, March 1939. WL/SA/FPA/A7/1; *Approved List*, March 1940. WL/SA/FPA/A7/1; *Approved List*, February 1942. WL/SA/FPA/A7/1.
95. NBCA Medical Sub-Committee Minutes, 16 May 1938. WL/SA/FPA/A5/88.
96. NBCA Medical Sub-Committee Minutes, 31 October 1938. WL/SA/FPA/A5/88.
97. NBCA Medical Sub-Committee Minutes, 30 November 1938. WL/SA/FPA/A5/88.
98. Borge, *Protective Practices*, 64; Löwy, '"Sexual Chemistry" before the Pill', 245–74, 255.
99. NBCA Medical Sub-Committee Minutes, 31 October 1938. WL/SA/FPA/A5/88.
100. Löwy, 'Sexual Chemistry', 256–7.
101. Letter R. Squire to A. Packard, 2 March 1938. Wellcome Library, Archives of the Rockefeller Family Archives and English Eugenics Society (WL/GC), WL/GC/88/1.
102. National Committee on Maternal Health Memorandum, 3 March 1938. WL/GC/88/1.
103. NBCA Medical Sub-Committee Minutes, 7 March 1939. WL/SA/FPA/A5/88.
104. For further discussion of British contraceptive manufacturing and marketing in the twentieth century see Borge's *Protective Practices* and Jones' *The Business of Birth Control*.
105. Letter C. Chance to M. Sanger, 13 March 1928. MSP-LC. Quoted in Borell, 'Birth Control Research', 82.
106. Borell, 'Birth Control Research', 81.
107. NBCA Circular, February 1938. WL/SA/FPA/A7/1.
108. Barbara Evans, *Freedom to Choose: The Life and Work of Dr Helena Wright, Pioneer of Contraception* (London: The Bodley Head, 1984), 133–4.

109. Birth Control Investigation Committee Promotional Pamphlet, 1931. Wellcome Library, Archives of the Private Papers of Margery Spring Rice (WL/SA/SR), WL/SA/SR19/1-5.
110. Kate Fisher, 'Contrasting Cultures of Contraception: Birth Control Clinics and the Working-Classes Between the Wars', in *Biographies of Remedies: Drugs, Medicines and Contraceptives in Dutch and Anglo-American Healing Cultures*, eds. Marijke Gijswijt-Hofstra, G. M. van Heteren and Tilli Tansey (Amsterdam: Rodopi, 2002), 146–7.
111. Draft Statement on the Work of the BCIC by C. P. Blacker and J. Huxley, 1930. WL/SA/FPA/A13/5; Report of Berlin Visit Investigating the Gräfenberg Ring Contraceptive, December 1929. WL/SA/FPA/A13/5; Caroline Rusterholz, 'Testing the Gräfenberg Ring in Interwar Britain: Norman Haire, Helena Wright and the Debate over Statistical Evidence, Side Effects and Intra-uterine Contraception', *Journal of the History of Medicine and Allied Sciences* 72, 4 (2017): 448–67.
112. Rose Holz, *The Birth Control Clinic in a Marketplace World* (Woodbridge: Boydell & Brewer, University of Rochester Press, 2012), 38–43.
113. Evelyn Fuller, *On the Management of a Birth Control Centre*, Second Edition (London: Noel Douglas, 1931), 30–1.
114. BCIC Memorandum on Work, 1931. WL/SA/FPA/A13/5.
115. Edith Howe-Martyn and Mary Breed, *The Birth Control Movement in England* (London: John Bale and Sons & Danielsson Ltd, 1930), 19.
116. Marie Stopes, *The First Five Thousand: Being the First Report of the First Birth Control Clinic in the British Empire* (London: John Bale and Sons & Danielsson Ltd, 1925).
117. Marie Stopes, *Equipping a Birth Control Clinic*, Second Edition (London: Mothers' Clinic for Constructive Birth Control, 1952), 9.
118. Stopes, *Equipping a Clinic*, 12.
119. Marie Carmichael Stopes, *Contraception (Birth Control): Its Theory, History and Practice* (London: Putnam & Co, Ltd, 1934), 447.
120. Stopes, *Contraception*, 178–9, 198–9, 448–9. In *The Business of Birth Control*, Jones discusses Stopes, her development of and dedication to the Pro-race Cap and battles with the medical establishment, legitimate and quackish contraceptive manufacturers and birth control competitors in greater depth.
121. Stopes, *First Five Thousand*, 14.
122. Stopes, *First Five Thousand*, 24.
123. Norman E. Himes and Vera C. Himes, 'Birth Control for the British Working Classes: A Study of the First Thousand Cases to Visit an English Birth Control Clinic', reprinted from *Hospital Social Service* 19 (1929): 596.

124. Ministry of Health, *Interim Report of Departmental Committee on Maternal Mortality and Morbidity* (London: His Majesty's Stationery Office, 1930), 44; Ministry of Health, *Final Report Departmental Committee on Maternal Mortality and Morbidity* (London: His Majesty's Stationery Office, Wyman and Sons, 1932), 130–1.
125. Norman E. Himes, 'British Birth Control Clinics: Some Results and Eugenic Aspect of Their Work', *Eugenics Review* 20, 3 (October 1928): 157–65.
126. Stopes, *The First Five Thousand*, 25.
127. Himes, 'British Birth Control Clinics', 159.
128. NKWWC Medical Committee Chairman's Report, October 1934. WL/SA/FPA/NK224.
129. Himes, 'British Birth Control Clinics', 162.
130. Stopes, *Contraception*, 13.
131. Howe Martyn and Breed, *The Birth Control Movement in England*, 17.
132. Himes, 'British Birth Control Clinics', 165.
133. Himes, 'British Birth Control Clinics', 164.
134. Himes, 'British Birth Control Clinics', 161.
135. Himes and Himes, 'Birth Control for the British Working Classes', 578–617.
136. Himes and Himes, 'Birth Control for the British Working Classes', 615.
137. Himes and Himes, 'Birth Control for the British Working Classes', 610.
138. Kate Fisher, *Birth Control, Sex and Marriage in Britain, 1918–1960* (Oxford: Oxford University Press, 2006), 131–3; Rusterholz, *Women's Medicine*, 65.
139. Himes and Himes, 'Birth Control for the British Working Classes', 611.
140. Himes and Himes, 'Birth Control for the British Working Classes', 612.
141. Fisher, *Birth Control, Sex and Marriage in Britain*, 131–3; Society for the Provision of Birth Control Clinics, *Birth Control and Public Health: A Report on Ten Years' Work of the Society for the Provision of Birth Control Clinics* (London, 1932).
142. Himes and Himes, 'Birth Control for the British Working Classes', 612–3.
143. Norman E. Himes, 'Some Untouched Birth Control Research Problems', *Eugenics* 3, 2 (1930): 3. WL/SA/SR15/24C.
144. Specification of Proposed Work for 12 & 14 Telford Road, 30 August 1932. WL/SA/FPA/NK/98.
145. NKWWC Medical Committee Chairman's Report, October 1934. WL/SA/FPA/NK224.
146. NKWWC Seventeenth Doctors Committee Report, 2 August 1934. WL/SA/FPA/NK225; Fuller, *Management of a Birth Control Centre*, 21–2.

147. NKWWC Eighteenth Medical Committee Report, 28 September 1934. WL/SA/FPA/NK225.
148. NKWWC Twentieth Medical Committee Report, 15 November 1934. WL/SA/FPA/NK225.
149. NKWWC Seventeenth Doctors Committee Report, 2 August 1934. WL/SA/FPA/NK224.
150. NKWWC Doctors Committee Report, [c.1935]. WL/SA/FPA/NK225.
151. Fuller, *Management of a Birth Control Centre*, 21–2.
152. Natasha Szuhan, 'The North Kensington Women's Welfare Centre Pre-war Patient Case Card', *Social History of Medicine*, forthcoming.
153. NKWWC Seventeenth Doctors Committee Report, 2 August 1934. WL/SA/FPA/NK225.
154. NBCA Medical Sub-Committee Minutes, 16 October 1937. WL/SA/FPA/A5/88.
155. NBCA Medical Sub-Committee Work during 1936, 1937. WL/SA/FPA/A5/88.
156. NBCA Medical Sub-Committee Minutes, 5 July 1936. WL/SA/FPA/A5/88; NBCA Medical Sub-Committee Minutes, 16 October 1937. WL/SA/FPA/A5/88.
157. Rusterholz, *Women's Medicine*, 65–8, 152–3.
158. NKWWC Medical Committee Minutes, 30 September 1958. WL/SA/FPA/NK230; NKWWC Medical Committee Minutes, 29 October 1959. WL/SA/FPA/NK230; NKWWC Medical Committee Minutes, 28 September 1953. WL/SA/FPA/NK230.
159. NKWWC Medical Committee Minutes, 3 April 1952. WL/SA/FPA/NK230; NKWWC Medical Committee Minutes, 5 January 1953. WL/SA/FPA/NK230.
160. NKWWC Medical Committee Minutes, 24 March 1960. WL/SA/FPA/NK230.
161. NKWWC Doctors Committee Notes, 6 July 1934. WL/SA/FPA/NK224.
162. NKWWC Medical Committee Report, 4 January 1935. WL/SA/FPA/NK224.
163. NKWWC Medical Committee Report, 4 February 1935. WL/SA/FPA/NK224; NKWWC Nineteenth Medical Committee Report, 24 October 1934. WL/SA/FPA/NK225; Dorothy Porter, *Health, Civilization and the State: A Political History of Public Health from Ancient to Modern Times* (London: Routledge, 2000), 213.
164. NKWWC Medical Committee Report, 4 January 1935. WL/SA/FPA/NK224.
165. NKWWC Medical Committee Minutes, 23 May 1935. WL/SA/FPA/NK225.

166. NKWWC Clinic Superintendent's Report, 22 April 1937. WL/SA/FPA/NK226; NKWWC Clinic Superintendent's Report, 17 November 1937. WL/SA/FPA/NK226.
167. NBCA Medical Sub-Committee Minutes, 16 October 1937. WL/SA/FPA/A5/88.
168. NBCA Ad Hoc Committee on Testing Minutes, 22 November 1937. WL/SA/FPA/A5/88.
169. Himes and Himes, 'Birth Control for the British Working Classes', 608.
170. NKWWC Clinic Superintendent's Report, 22 April 1937. WL/SA/FPA/NK226.
171. NKWWC Clinic Superintendent's Report, 17 November 1937. WL/SA/FPA/NK226.
172. Fuller, *Management of a Birth Control Centre*, 24.
173. Andrea Tone, 'Contraceptive Consumers: Gender and the Political Economy of Birth Control in the 1930s', *Journal of Social History* 29, 3 (1996): 492–3.
174. NBCA Medical Sub-Committee Minutes, 17 October 1936. WL/SA/FPA/A5/88.
175. NBCA Medical Sub-Committee Minutes, 17 October 1936. WL/SA/FPA/A5/88; Szuhan, 'Sex in the Laboratory', 500.
176. NKWWC Medical Committee Minutes, 30 November 1936. WL/SA/FPA/NK226.
177. Rusterholz, *Women's Medicine*, 65.
178. Appendix to NBCA Medical Sub-Committee Minutes, 24 April 1937. WL/SA/EUG/D/12/16/22.
179. NKWWC Medical Committee Minutes, 15 November 1934. WL/SA/FPA/NK225.
180. NKWWC Medical Committee Minutes, 15 November 1934. WL/SA/FPA/NK225.
181. NBCA Medical Sub-Committee Minutes, 12 June 1939. WL/SA/FPA/A5/88.
182. NBCA Medical Sub-Committee Minutes, 12 June 1939. WL/SA/FPA/A5/88.
183. NBCA Medical Sub-Committee Minutes, 7 March 1939. WL/SA/EUG/D/12/16/22.
184. NBCA Medical Sub-Committee Minutes, 7 March 1939. WL/SA/EUG/D/12/16/22.
185. Michael Fielding, *Practical Advice on Birth Control* (London: National Birth Control Association, 1933).

    WL/SA/SP17/12/17; Jones, *The Business of Birth Control*, 120; Szuhan, 'Sex in the Laboratory', 500.
186. NBCA Medical Sub-Committee Minutes, 5 May 1935. WL/SA/EUG/D/12/16/22.

187. NKWWC Medical Committee Minutes, 20 February 1941. WL/SA/FPA/NK226.
188. NKWWC Medical Committee Minutes, 3 December 1941. WL/SA/FPA/NK226.
189. NBCA Medical Sub-Committee Minutes, 18 May 1941. WL/SA/FPA/A5/88.
190. NBCA Medical Sub-Committee Minutes, 18 May 1941. WL/SA/FPA/A5/88.
191. NBCA Medical Sub-Committee Minutes, 7 June 1942. WL/SA/FPA/A5/88.
192. NBCA Medical Sub-Committee Minutes, 30 March 1943. WL/SA/FPA/A5/88; NKWWC Medical Committee Minutes, 18 February 1947. WL/SA/FPA/NK227.
193. NBCA Medical Sub-Committee Minutes, 7 June 1942. WL/SA/FPA/A5/88.
194. NBCA Medical Sub-Committee Minutes, 7 March 1939. WL/SA/FPA/A5/88.
195. NBCA Medical Sub-Committee Minutes, 20 May 1939. WL/SA/FPA/A5/88.
196. NBCA Medical Sub-Committee Minute, 27 October 1947. WL/SA/FPA/A5/88; NKWWC Medical Committee Minutes, 19 May 1948. WL/SA/FPA/NK227; NBCA Medical Sub-Committee Minutes, 28 January 1947. WL/SA/FPA/A5/88.
197. Helena Wright, *Contraceptive Technique: A Handbook for Medical Practitioners and Senior Students* (London: J&A Churchill Ltd., 1951), 58.
198. NKWWC Medical Committee Minutes, 30 December 1950. WL/SA/FPA/NK227.
199. NKWWC Medical Committee Minutes, 22 May 1951. WL/SA/FPA/NK227.
200. NKWWC Medical Committee Minutes, 15 October 1954. WL/SA/FPA/NK230; NKWWC Medical Committee Minutes, 26 November 1956. WL/SA/FPA/NK230.
201. NKWWC Medical Committee Minutes, 24 March 1960. WL/SA/FPA/NK230.
202. NKWWC Medical Committee Minutes, 17 September 1964. WL/SA/FPA/NK230.
203. NBCA Medical Sub-Committee Minutes, 5 July 1936. WL/SA/FPA/A5/88.
204. NBCA Medical Sub-Committee Minutes, 5 July 1936. WL/SA/FPA/A5/88.
205. NBCA Medical Sub-Committee Minutes, 17 October 1936. WL/SA/FPA/A5/88; NBCA Medical Sub-Committee Minutes, 5 July 1936. WL/SA/FPA/A5/88.

206. NBCA Medical Sub-Committee Minutes, 31 January 1938. WL/SA/EUG/D/12/16/22.
207. NBCA Medical Sub-Committee Minutes, 24 April 1937. WL/SA/FPA/A5/88; NBCA Ad Hoc Committee on Testing Minutes, 22 November 1937. WL/SA/FPA/A5/88.
208. NKWWC Medical Committee Minutes, 23 January 1962. WL/SA/FPA/NK230.
209. NKWWC Medical Committee Minutes, 25 March 1965. WL/SA/FPA/NK230.
210. Letter C. Harvey to L. Arnold, 21 June 1950. WL/SA/FPA/A7/13.1.
211. FPA List of Approved Contraceptives Report, 1953. WL/SA/FPA/A7/16.
212. FPA List of Approved Contraceptives Report, 1953. WL/SA/FPA/A7/16.
213. Graeme Gooday, 'Placing or Replacing the Laboratory in the History of Science?', *Isis* 99, 4 (2008): 788.
214. N.W. Pirie, 'Biochemistry of Conception Control', *Eugenics Review* xliv, 3 (1952): 139.
215. NBCA Circular, February 1938. WL/SA/FPA/A7/1.
216. "Birth Control', *The Lancet* 216, 5577 (1930):147; Contraceptives: Effect on Fertility, 1943–9. WL/SA/FPA/A7/16A. This folder is dedicated to research requests and letters regarding contraceptives causing sterility and malformation.
217. H. A. R. Binney Meeting Minutes, 10 January 1955. WL/SA/FPA/A7/20; FPA List of Approved Contraceptives Report, 1953. WL/SA/FPA/A7/16.
218. Letter FPA General Secretary to P. Schidrowitz, 4 February 1937. WL/SA/FPA/A7/20.
219. BCIC Draft Statement of Work, 1930. WL/SA/FPA/A13/5.
220. Memorandum on Harmlessness Tests for the Eugenics Society, October 1954. WL/SA/FPA/A7/16.
221. Letter C. Harvey to S. C. S. Robinson, 2 October 1948. WL/SA/FPA/A7/16.
222. BCIC Draft Statement of Work, 1930. WL/SA/FPA/A13/5.
223. NBCA Medical Sub-Committee Minutes, Session One, 17 October 1936. WL/SA/FPA/A5/88.
224. NBCA Medical Sub-Committee Minutes, 6 October 1935. WL/SA/FPA/A5/88.
225. NBCA Medical Sub-Committee Minutes, 26 January 1936. WL/SA/FPA/A5/88.
226. NBCA Medical Sub-Committee Work during 1936, 1937. WL/SA/FPA/A5/88.

227. NBCA Medical Sub-Committee Minutes, 5 July 1936. WL/SA/FPA/A5/88.
228. Memorandum on Chemical Contraceptives Tested by C. Harvey, 22 June 1939–October 1942. WL/SA/FPA/A7/16.
229. Letter Unknown FPA Member to R. W. Vernes, 4 October 1940. WL/SA/FPA/A7/1.
230. Notes on Approved List of Contraceptives, 23 March 1953. WL/SA/FPA/A7/16.
231. H. A. Davidson, *The Contact Test: A Method of Measuring Spermicidal Action* (London: The Lancet Office, 1953), 2–3. WL/SA/FPA/A7/16.
232. Jones, *The Business of Birth Control*, 147–59.
233. Interim Statement on the Statistical Implications of the Approved List of 1952, 19 January 1953. WL/SA/FPA/A7/3. Even Volpar couldn't pass the Diffusion Test, yet it remained on the approved list Section A.
234. H. A. Davidson Diffusion Test Notes, 20 December 1951. WL/SA/FPA/A7/7; H. A. Davidson Diffusion Test Notes, 19 February 1952. WL/SA/FPA/A7/7.
235. H. A. Davidson, 'The Contact Test: A Method of Measuring Spermicidal Action', *Lancet* 262, 6783 (1953): 432–4.
236. Letter C. Harvey to S. C. S. Robinson, 2 October 1948. WL/SA/FPA/A7/16.
237. Letter C. Harvey to I. James, 2 January 1952. WL/SA/FPA/A7/16.
238. Letter FPA General Secretary to C. Harvey, 14 January 1952. WL/SA/FPA/A7/16.
239. Memorandum on Harmlessness Tests for the Eugenics Society, October 1954. WL/SA/FPA/A7/16.
240. Letter FPA General Secretary to P. Eckstein, 5 June 1953. WL/SA/FPA/A7/16.
241. Letter P. Eckstein to I. James, 7 November 1953. WL/SA/FPA/A7/16.
242. Letter M. Jackson to P. Eckstein, 12 November 1953. WL/SA/FPA/A7/16.
243. Telephone note from M. Jackson, 28 April 1954. WL/SA/FPA/A7/16.
244. Memorandum on Harmlessness Tests for the Eugenic Society, October 1954. WL/SA/FPA/A7/16.
245. Letter I. James to P. Eckstein, 29 April 1954. WL/SA/FPA/A7/16; Telephone note from M. Jackson, 28 April 1954. WL/SA/FPA/A7/16.
246. Results of a New Series of Harmlessness Testing on Monkeys, 8 October 1955. WL/SA/FPA/A7/16.
247. Memorandum on Harmlessness Tests for the Eugenics Society, October 1954. WL/SA/FPA/A7/16.
248. Draft Letter I. James to P. Eckstein, October 1954. WL/SA/FPA/A7/16. Szuhan, 'Sex in the Laboratory', 507.

249. Results of Harmlessness Tests in Monkeys, November 1954. WL/SA/FPA/A7/16.
250. Results of a New Series of Harmlessness Testing on Monkeys, 8 October 1955. WL/SA/FPA/A7/16.
251. Letter E. Mears to M. Jackson, 30 October 1959. WL/SA/FPA/A7/16.
252. International Planned Parenthood Federation (IPPF) Fourth Meeting of Sub-Committee on Tests for Contraceptive Products, February 1959. WL/SA/FPA/A10/15.
253. IPPF Second Meeting of Sub-Committee on Tests for Contraceptive Products, 14 October 1957. WL/SA/FPA/A10/15.
254. NBCA Medical Sub-Committee Minutes, Session Two, 28 July 1934. WL/SA/FPA/A5/88.
255. NBCA Medical Sub-Committee Minutes, Session One, 3 March 1935. WL/SA/FPA/A5/88; NBCA Medical Sub-Committee Minutes, Session Two, 3 March 1935. WL/SA/FPA/A5/88; NBCA Medical Sub-Committee Minutes, Session One, 4 May 1935. WL/SA/FPA/A5/88; NBCA Medical Sub-Committee Minutes, Session One. 21 July 1935. WL/SA/FPA/A5/88.
256. Letter P. Schidrowitz to FPA General Secretary, 12 June 1939. WL/SA/FPA/A7/20.
257. Slaton, '"As Near as Practicable"', 58–60; Szuhan, 'Sex in the Laboratory', 507.
258. NBCA Medical Sub-Committee and Ad Hoc Sub-Committee Minutes, 20 May 1937. WL/SA/FPA/A5/88.
259. Letter FPA General Secretary to P. Schidrowitz, 4 February 1937. WL/SA/FPA/A7/20; Memorandum from P. Schidrowitz, 22 February 1937. WL/SA/FPA/A7/20.
260. Letter P. Schidrowitz to FPA Secretary, 14 January 1938. WL/SA/FPA/A7/20.
261. Letter H. Holland to P. Schidrowitz, 30 October 1941. WL/SA/FPA/A7/20; Letter P. Schidrowitz to S. C. S. Robinson, 18 October 1948. WL/SA/FPA/A7/20; Results of "Dutch Cap" Test by P. Schidrowitz, 28 April 1949. WL/SA/FPA/A7/20; Szuhan, 'Sex in the Laboratory', 507–8.
262. Memorandum from P. Schidrowitz, 22 February 1937. WL/SA/FPA/A7/20.
263. NBCA Medical Sub-Committee Minutes, 16 October 1935. WL/SA/FPA/A5/88.
264. NBCA Medical Sub-Committee and Ad Hoc Sub-Committee Minutes, 22 November 1937. WL/SA/FPA/A5/88.
265. NBCA Medical Sub-Committee Minutes, 31 January 1938. WL/SA/FPA/A5/88.

266. NBCA Medical Sub-Committee and Ad Hoc Sub-Committee Minutes, 22 November 1937. WL/SA/FPA/A5/88.
267. NBCA Medical Sub-Committee Minutes, 31 January 1938. WL/SA/FPA/A5/88.
268. Letter FPA Acting Secretary to P. Schidrowitz, 17 October 1941. WL/SA/FPA/A7/20.
269. Results of 'Lambutt' Cap Test by P. Schidrowitz, 30 April 1942. WL/SA/FPA/A7/20.
270. FPA Medical Sub-Committee Minutes, 20 May 1940. WL/SA/FPA/A7/1.
271. Borge, *Protective Practices*, 99–102.
272. Szuhan, 'Sex in the Laboratory', 508.
273. Letter R. W. Vemes to H. Holland, 3 April 1941. WL/SA/FPA/A7/1. Their emphasis.
274. Borge, *Protective Practices*, 100–2; Jones, *The Business of Birth Control*, 42–5, 148–9.
275. Letter FPA General Secretary to A. R. Reid, 18 March 1954. WL/SA/FPA/A7/20.
276. Letter R. Edwards to I. James, 17 October 1953. WL/SA/FPA/A7/20.
277. Letter Reid to FPA, 29 August 1952, WL/SA/FPA/A7/73. Quoted in Borge, *Protective Practices*, 77.
278. Interview with R. Edwards, 10 November 1953. WL/SA/FPA/A7/20.
279. *Approved List of Contraceptives: Specifications of Tests &c. of Rubber*, August 1953. WL/SA/FPA/A7/20; Szuhan, 'Sex in the Laboratory', 509.
280. Borge, *Protective Practices*, 95.
281. Borge, *Protective Practices*, 95.
282. Letter FPA General Secretary to A. R. Reid, 14 October 1955. WL/SA/FPA/A7/20. In 1935 the medical sub-committee found Prentif Dumas cap sizes unsatisfactory and to standardise size, asked that 'Messer's Prentif possess themselves of a set of the ordinary Dumas Caps sizes small, medium and large, as made by Lamberts, and use them as their standard'. NBCA Medical Sub-Committee Minutes, 13 January 1935. WL/SA/FPA/A5/88.
283. Telephone Message from I. James, 8 January 1954. WL/SA/FPA/A7/20.
284. Test for Contraceptives on the Approved List, 25 October 1954. WL/SA/FPA/A7/20.
285. Ernest Dichter Associates, 1739e: Some Remarks on the Questions Raised on the Contraceptive Research (London: Ernest Dichter Associates, 1961), 1. Box 78, Ernest Dichter Papers, Hagley Museum and Library, Wilmington, De 19807. Quoted in Borge, *Protective Practices*, 96.

286. Letter FPA General Secretary to P. M. C. Watkins, 5 January 1954. WL/SA/FPA/A7/20; Revised Insert from Lamberts, February 1954. WL/SA/FPA/A7/20.
287. Response Lamberts to FPA, 2 February 1954. WL/SA/FPA/A7/20.
288. Letter The Mother Welfare Clinic Secretary to I. James, 9 February 1953. WL/SA/FPA/A7/20.
289. Letter Lamberts to I. James, 4 November 1953. WL/SA/FPA/A7/20.
290. Borge, *Protective Practices*, 94–6.
291. *Approved List*, 1958. WL/SA/FPA/A7/5.

CHAPTER 5

Contraceptive Standards in the Age of the Pill: Influencing and Exporting Formal Oversight

Abstract This chapter charts the Family Planning Association's pill-era achievements to implement broader influence and receive state legitimisation through collaborations with regulators like the British Standards Institution, British Pharmacopoeia and British Pharmacopoeia Commission, and Council for the investigation of Fertility Control, and less formally the medical profession, as they were all slowly persuaded to overtake formal contraceptive regulatory oversight. It further charts the socio-cultural and medico-legal challenges the association faced as the contraceptive landscape began to fundamentally shift in the 1950s and 1960s with increased acceptability, and the emergence and introduction of the hormonal oral contraceptive pill. In this period the association moved out of the shadows and into greater public focus, built its British empire through the expansion of its sites and attempted to broaden its socio-political influence through a series of partnerships and collaborations, funding opportunities and further efforts to streamline and guarantee its approved methods and products, and those emerging, through an extended research agenda.

This chapter charts the big contraceptive, socio-cultural and medico-legal challenges the Family Planning Association (FPA) faced as the contraceptive landscape began to fundamentally shift in the 1950s and 1960s with the increased acceptability of the practice and the emergence and

N. Szuhan, *The Family Planning Association and Contraceptive Science and Technology in Mid-Twentieth-Century Britain*, Medicine and Biomedical Sciences in Modern History, https://doi.org/10.1007/978-3-030-81300-0_5

introduction of the hormonal oral contraceptive pill. This was a boom period for the association, manufacturers, contraceptive technologies and widespread official and unofficial support and demand for safe, effective, acceptable and reasonably priced products. In this period the FPA moved out of the shadows and into greater public focus, built its British empire through the expansion of its sites, and attempted to broaden its socio-political influence through a series of partnerships and collaborations, funding opportunities and further efforts to streamline and guarantee its approved methods and products, and those emerging, through an extended research agenda.

The advent of hormonal contraceptives in the 1950s, and their abrupt mass appeal prompted a major regulatory transition in Britain. The government and medical profession finally grasped the association's long held position that an unofficial body should not be responsible for contraceptive standardisation and oversight. As oral contraceptives were limited to clinical prescription, they became directly and consistently synonymous with medicine. The shocked profession was forced to admit the pill was a contraceptive uniquely within its responsibility and control, irrespective of patient need or desire.

The association's long-standing pure and applied research programs were brought to the fore in its legitimisation and standardisation efforts as it sought to attract the interest of rubber and pharmaceutical watchdogs to overtake its ad-hoc regulatory work; and support the medical profession in effecting methods for oral contraceptive oversight, both through safety and efficacy testing at FPA sites and working with contraceptive testing satellite the Council for the Investigation of Fertility Control (CIFC). Through these contacts the association was slowly but finally able to achieve a transfer of its regulatory authority and to help shape the standards that would be applied by government-recognised bodies. I argue that by expanding its medico-scientific research programs to include harm and ensure quality, the NBCA/FPA was increasingly able to convince British supervisory bodies that contraception was so widely used and accepted, it warranted official oversight by regulators like the British Standards Institution (BSI) the Royal Pharmacopoeia Society through the British Pharmacopoeia Commission and the broader medical profession.

Exporting the Approved List and Contraceptive Standards Research to Official Regulators

The association had no intention to define and oversee contraceptives in perpetuity. It always planned to abdicate this responsibility as soon as an opportunity presented itself. The FPA bided its time until the mid-twentieth century when social, political and technological developments rendered contraception a matter of both personal and public health. Tanya Evans argues that before the 1960s 'knowledge and technology played very little part in the sex lives of most people' but concedes that many other historians 'doubt the veracity of individuals' claims to profound sexual ignorance'.[1] FPA records demonstrate that the truth lies somewhere between these divergent interpretations. People knew of and practiced contraception liberally, but the topic rarely received positive press or public discussion. The social history of contraception is a brilliant example of the fact that sex and related topics were historically 'represented rather than experienced, [and] controlled rather than enjoyed'.[2] Therefore, in light of these contrasting cultures of secrecy and silence and increased contraceptive familiarity and practice, some liberalisation of attitudes towards sex and its social, familial and moral consequences had to precede formal moves towards contraceptive legitimisation through official regulatory channels and publications. As was discussed in Chapter 3, there were cultural changes that emerged in the post-war period which evinced changes in sexual mores and ideals. The association's avant-garde sexual, medical and educational female focus proved prophetic as women became central to the emerging progressive British sexual and contraceptive cultures. Observer of contemporary sexual culture, Eustace Chesser, explained that in the immediate aftermath of the war, women's sexual happiness and satisfaction in a companionate marriage became one of the central markers of marital and family formation, structure and success.[3] In the post-sexual revolution West, this would necessarily refer to female climax or orgasm, but as Szreter and Fisher's oral histories have proved, in the 1940s and 1950s female sexual satisfaction was also intimately linked to contemporary ideals of gender performance, which included being a good wife and responsible mother by performing traditional women's sexual and reproductive duties.[4] So, there was delicately balanced and slow-moving sexual liberalisation that somewhat equalised gendered sexuality and decentred parenthood that took place in individual bedrooms as well as in sites of social, medical and educational sexual advocacy. Thus, it would not be quite correct to declare that huge public and private sphere changes had taken place regarding openness about sex and contraception, but rather that significant, but personal and liberal small, generational conceptual and practical ones had been effected at discreet levels that led to demographic, educational and commercial changes linked to contraception; and that the FPA effectively charted and played upon these to petition regulators to take an interest in and ultimately regulate the safety, efficacy and acceptability of contraceptive products. Its two initial targets were the British Pharmacopoeia Commission that set 'authoritative official standard[s] for pharmaceutical substances and medicinal products', and the BSI that regulated production and quality of merchandise manufactured and sold in Britain. The association's interaction and collaboration with the BSI to set rubber standards is discussed below, with the Pharmacopoeia Commission included where applicable.

The association first attracted the attention and encouragement of regulators in 1953, following a 'chance' encounter between the clinics sub-committee secretary Patricia Cripps, and Professor Ronald S. Edwards, a London School of Economics industrial administration and regulation specialist. Cripps memorialised the meeting as Edwards declared he knew of the approved list and expressed keenness to help the FPA make contacts and gain 'publicity in a sphere in which [it was] at present unknown'.[5] This serendipity offered an invaluable prospect for the association to redistribute its regulatory responsibility.

Edwards explained that his eagerness to make contact was because unlike most independent regulatory 'schemes [that] run into difficulty of one kind or another', the approved list had proved remarkably 'successful'.[6] He keenly queried the association's engagement with manufacturers, specifications and standards, how it ensured ongoing quality, and promoted the approved list. General Secretary, Irene James, responded that it was 'still in the process of getting some of the answers' and pressed Edwards for a meeting on the matter.[7] They met within weeks, and Edwards informed a delighted James that he had in the interim approached 'his friend Mr [H. A.R] Binney (now Director of British Standards Institution)' who had indicated a willingness to assume the 'administrative support side of standardisation (and factory inspection) of rubber goods'; but refused to assume contraceptive regulation policy 'owing to the [institution's] Roman Catholic element'. However, Edwards advised it was unlikely the religious constituents would object if the FPA begged for assistance.[8] So the association leaders arranged to make Binney's acquaintance, to foster the relationship and extend their regulatory agenda.

Buoyed by this support, the association also petitioned and enjoyed a 'sympathetic and encouraging response' from the British Pharmacopoeia Commission. That body had avoided including spermicidal pessaries, creams and gels in the *British Pharmacopoeia*, but it was hoped that in the wake of the Royal Commission on Population and the foundation of the National Health Service (NHS), and moves to obtain government funds for contraceptive work, that some the editors might be willing to help legitimise contraceptive efficacy and harmlessness standards. Hoping for spermicides to be included as early as 1954, the association submitted all the contraceptive compounds featured in the approved list's Section A. But the commission was not yet ready, and to delay having to action the request, editors Thomas C. Denstone and K. R. Capper requested the FPA submit a 'statement of the background, reasons etc. for [their] application for B.P. recognition', 'tentative monographs' on each type of compound, and an account of the FPA's testing and approval processes to date.[9]

As Binney and Edwards had 'suspected',[10] upon being presented with proposals for a transfer of regulatory power, both the BSI and British Pharmacopoeia executive committees were persuaded their administration had some duty to safeguard contraceptives.[11] Both supported the Royal Commission on Population's findings that individual couples being able to effectively control the size and time of their family was 'one of the first conditions of their own and the community's welfare', and that contemporary 'mechanical and chemical methods of contraception' must be accepted as viable tools to achieve those ends.[12]

The Royal Commission shaped encouraging community and regulatory attitudes toward contraception that the association quickly moved to exploit. The FPA provided evidence to the Commission based on its expertise and experience. It also exploited the Royal Commission to 'advocate and promote the provision of facilities for scientific contraception' and push for its contraceptive and fertility work to be recognised as an integral 'part of the public health service'.[13] The association advertised its medico-scientific credentials and partnerships to shine a light on its work with 'married couples', providing assistance in all areas of wedded and family difficulty; including but not limited to 'contraception, subfertility, gynaecological ailments, marital disharmony and psychological difficulties'.

Its submissions to the Royal Commission specifically included an appendix detailing FPA efforts to assist couples to address the spectrum of fertility concerns. This detailed the association's contraceptive as well as sub-fertility research and its affiliation with the medical community that often deferred to FPA physicians' expertise when fertility and contraceptive issues emerged in practice.[14] Two FPA representatives gave evidence to the commissioners, association secretary Margaret Pyke and long-term clinic physician and sexual health and fertility researcher Joan Malleson. They detailed the association's pure and applied research efforts and clarified that the FPA was dedicated to both preventing unwanted pregnancies and facilitating desired births.[15] The association had by then founded 'between 50 and 60 branch clinics' and each treated about '20 new patients at each session and [held] three sessions a week'; therefore 'between one thousand and fifteen hundred' new patients attended each site annually, in addition to four thousand returnees. Very small or rural clinics treated 'perhaps 120 patients per year'.[16] Together Pyke and Malleson chronicled the FPA's vital public- and nation-focused not-for-profit work for the commissioners, and offered unwavering assertions and evidence that association clinics provided an appropriate venue to address contraceptive and fertility issues within scientific structures and provided apt treatment from specialist physicians. The women imbued the association's work with the scientific and medical authority it always strived to project.

Throughout Malleson and Pyke's testimony, science was skilfully represented as both the master of contraception and the last hope of the infertile. Therefore, questioning its legitimacy in contraceptive and reproductive medical research was almost blasphemous. In the post-war era in which William Osler's prediction that 'the future belongs to science' was becoming more overtly true daily, the FPA was effectively using scientific evidence and rhetoric to demonstrate that reproductive science and technology belonged as much in the bedroom as the hospital, laboratory, factory and farm.[17] By 1950, scientific knowledge and expertise was guiding other aspects of daily life—work, health warfare, food, production, etc.—and its adaptation into an action, method or field that was otherwise questionable elevated the status of that subject to match any other supported by scientific evidence. The FPA always held that contraceptive medicine and science were congruent with scientific schools like biochemistry, neurology, endocrinology, etc., that were employed to understand and heal the human body. The association's leadership now perceived it was making a persuasive argument for its research agenda to a receptive professional audience whose backing would help to foster publicity and support for science-based contraceptive standards.

From the 1950s increasingly invasive mechanical and medical reproductive technologies (such as FPA physician and researcher Margaret Jackson's work to provide 'artificial insemination' services to infertile couples) were devised, trialled and applied by scientists, physicians and consumers.[18] This led some commonly utilised and less confronting contraceptive methods like condoms, spermicides and diaphragms/caps, to be normalised and destigmatised. Further, in that decade a church and 'medically approved' product was marketed. The Conception Day Indicator was claimed to be based on the scientific assessment of the menstrual cycles' 'fertile and infertile' days to provide women a means to track the 'safe period' for intercourse.[19] The FPA vehemently opposed the product as its own research into the rhythm method's efficacy had proved it was only applicable if a 'woman's cycle is sufficiently regular and maximum safety is not required'.[20] Although the product operated in stark contrast to the association's long-standing dual contraceptive prescription for guaranteed safety, the implicit approval from the church for scientific contraception signposted by this product, denoted a cultural shift and promotional opportunity that was too valuable for the association to miss.[21] The Royal Commission's investigation, the advent of a religiously sanctioned contraceptive method, and the advance of modern medical reproductive programs each demonstrated that prohibitions surrounding sex and contraception were subsiding. It was now clear that contraceptive regulation would be overtaken by official bodies, only when and how remained to be resolved. Gaining broad medical, religious and popular support for the wide dissemination of scientific contraceptive knowledge and methods, filled the FPA with every confidence of a similarly positive response from regulators. Early contact with both the BSI and British Pharmacopoeia Commission proved overwhelmingly positive, and happily only one well-timed and tenacious petition to each proved necessary.

Although both the BSI and British Pharmacopoeia Commission agreed contraception was appropriate for them to oversee in principle, neither handover was straightforward. Each formed technical committees, comprised of experienced members of other regulatory boards, manufacturers' representatives and the FPA, to guide and oversee contraceptive regulatory frameworks from concept to publication and implementation.[22] The committee was keen to safeguard a 'much closer control of quality and effectiveness of contraceptives' than the FPA was able to achieve through its cooperative regulatory efforts.[23] As the association was already entrenched in this work, it was charged with ensuring (majority) manufacturers' support for both the BSI and British Pharmacopoeia projects, drafting the standard/monograph for each, and formulating foolproof, up-to-date scientific tests to assess products pre- and post-approval.[24] Both regulators were ready to undertake contraceptive oversight, providing an external organisation, ideally the FPA, performed any necessary controversial work. To ensure this transition of power, the association agreed to this deal.

Whilst support was generally forthcoming, some manufacturers had to be persuaded of the value of formalised contraceptive regulation. Upon learning of the BSI's plan to establish a rubber contraceptive standard, the Federation of British Rubber Manufacturers Association declared 'no standard was necessary, as the public were perfectly satisfied' with the quality of available products.[25] However, A. R. Reid of London Rubber Company, which dominated the mid-twentieth century British rubber contraceptive market, producing over 80 per cent of prophylactics,[26] 'wanted to co-operate with the FPA' agenda. The upshot being that even without any other manufacturers support, the FPA quickly and easily met the BSI request that the 'principal manufacturers... agree with the scheme'.[27] However, the long-running contention between manufacturers and the FPA regarding the dating of rubber goods threatened to quash this alliance, as London Rubber remained the company most at odds with the principle of dating into the 1950s.[28] This matter was first raised twenty years earlier when dating became a prime condition for approved list endorsement, and was also integral to the BSI deal as the current British Standard for Rubber Surgical Goods included an expiry date.[29] In the FPA's first revision of its approved list rubber testing procedures following BSI contact, it specified all tests were 'carried out as far as possible in accordance with British Standard Specification 903'[30] testing standards and methods. To this end, the FPA was adamant that dating was vital and demanded 'all articles [had] to be marked "Use before..." or "To be used before..." with the relevant date'.[31]

The prominence of dating and the optimum lifespan of some approved list rubber appliances came into question in the 1950s. Some clinics reported that despite being clearly dated, some products' lifespan in storage or use was significantly shorter than estimated. London Rubber became embroiled in a dating controversy in early 1954 when it emerged that 'through a misunderstanding', the dates which had previously been printed on 'control slips' of Paragons washable sheaths had been 'discontinued'. This reopened long-standing disagreements between the association and business that had dissipated since they had come to mutually beneficial cooperative agreements regarding advertising and product costs. FPA Secretary Irene James chided the company that it was in breach of approved list requirements that protectives must contain a 'packing slip showing the final date' of reliable use.[32] The dominant manufacturer, and indeed leading clinic supplier, retorted that date slips were 'pointless for Paragons', as they were made from a 'much heavier material [than condoms, that was] practically indestructible, as regards storage life'.[33] The company objected to the FPA and thus the BSI's mandatory dating policy on several grounds. First, it manufactured and sold 'from three quarters to one million condoms every week' and was 'responsible for four-fifths of the total consumption in the country', so sales would be impacted if consumers were dissatisfied with the quality. Further, it refused 'to engage the large numbers of extra staff which [mandatory] dating would involve', especially as its 'products were made of latex, which did not perish' and their stock turnover was consistently high. This proved its products would not be 'kept in stock' long enough to notably degrade.[34] Finally, and confusingly given the previous arguments, it contended its 'protectives do store perfectly in any climate for at least seven years', and further claimed to have 15-year-old prophylactics that 'remained just as good as on the day they were made'. Despite its protests, the matter was again resolved by London Rubber agreeing to include a dated 'packing slip' in every pack. But it's in-principle objection to dating was ongoing.[35] As other historians have shown, the relationship between the pair was never outright acrimonious, but always teetered on the brink of conflict, and was always resolved in ways that offered various levels of mutual benefit.[36] In this instance maintaining support from the expanding FPA made good business sense, and not pressing the prophylactic supplier too hard, didn't risk losing the financial benefits of cheap or free merchandise for distribution.

The likely source of this continuing disagreement was an early miscommunication over dating which arose when London Rubber first agreed to include an expiry date 'three years in advance of the date of dispatch'. The manufacturer shadowed this concession stating we 'trust the time will come when [the association] will feel that this dating process can be dispensed with altogether'. In fact, all the manufacturers engaging with the FPA testing and approval procedures considered the arrangement to be a meeting of equals with shared goals. Only with manufacturers' active support could the association effect any standards and so the FPA trod a fine line between formal regulation by directive and encouraging manufacturers' cooperation. London Rubber's position that its regulatory relationship with the FPA was based on 'co-operation and compromise' is an accurate description of the pre-1960s regulatory situation in Britain.[37] In 1964 when the first contraceptive rubber standard was launched by the BSI, the FPA's insistence on dating proved a great marketing ploy for London Rubber. The manufacturer was able to cite this dating as a guarantee, as well as a use by date after which unused products must be discarded and new ones purchased.[38] Thus, the demand was appropriated to improve sales as well as further enhance the company's already huge share of the market.

Rubber quality and durability failures were an industry-wide issue. In the early 1950s, the FPA discovered a Lambert's 'Eclipse Paragon' sheath that was so deteriorated it 'split when stretched'. This was a huge concern as malleability was obviously central to the everyday use and cleaning of sheaths. Upon inspection by the George Carruthers' research laboratory it was observed that the rubber had 'perished' and further that 'the paper in contact with the sheath [during its transportation had] become discoloured, presumably due to the chemicals used in the curing process'.[39] The sub-par item was returned to Lamberts with a letter explaining the FPA 'could establish at least part of [that batch's] age' due to its own excellent record keeping that showed it was less than three years old—a date that was not marked on its packaging. The FPA all but accused Lamberts of intentionally omitting the date and demanded a guarantee the company would forthwith 'comply with the [approved list dating] conditions'; especially as the 'anticipated life' of a sheath, condom, diaphragm or cap was imperative to guarantee their mutual aims to provide effective contraceptives.[40] It was to prevent unnecessary risks such as this from causing product failures that the association had implemented and enforced this condition and used the only major threat at its disposal to try to enforce the agreed rules when necessary. Lamberts well aware that the weight of the over-the-counter contraceptive market already far outweighed the FPA buying power, derided the threat to remove the products from the list and clinics and argued that the sheath in question had been produced and shipped to the clinic 'over two and a half years ago'. Lamberts preached that as products 'tend to deteriorate more quickly on the stock shelf', they should be 'use[d] within a year' and added that if 'harmful chemical contraceptives or greasy substances' were avoided, sheaths should have a life span of about nine months. This last claim showed Lambert's declarations of its merchandise's longevity to be grossly disingenuous. But in this politically sensitive period when the maintenance of good relations between the association and industrialists was imperative to achieving the desired regulatory handover, the FPA chose to accept Lambert's assurance that the Paragon Eclipse sheath was 'now very limited' in production and its lack of dating had been an oversight.[41] This was likely also because a year earlier, the pair had 'battle[d]' over testing, dating and packaging and Lamberts had evidenced its long history of exemplary adherence to the approved list rules and regulations. Lamberts had further chastened the association for failing to deliver current information on enhanced approved list testing procedures and seeming to ascribe a more malleable dating standard for some other manufacturers. In 1954 Prentif also complained that it was 'complying with the approved list' regarding dating 'whilst other manufacturers products do not'.[42] The FPA was then on notice that companies that faithfully complied with its dating requirements were peeved that it was more indulgent with London Rubber about uniform dating. But despite this ire, as long as the FPA continued to buy, sell, promote and approve their goods in accordance with the entrenched quid pro quo, this contradiction didn't really impact or alter the complaining companies' relationship with the association. Manufacturers were further disinclined to make too much fuss as the projected regulatory transfer promised to significantly increase public exposure and thus sales by further legitimising contraception in the increasingly thriving mainstream consumer market.[43]

The association and all its regulatory and legitimisation ventures were tolerated by the industry as the obvious benefits of FPA success were to be joint gains, and any failures could easily be disavowed. In light of all of this, the association was especially eager to pass the regulatory baton as its initial approach to standardisation which championed 'co-operation and compromise' was effective in the age of legitimisation when the contraceptive sales and publicity quid pro quo effectively served both parties, but ultimately if the negotiated standards were not enforceable, and they were not, there was limited value in defining or implementing broader regulatory ideals. Thus, the realisation of its goal to handover regulation to the BSI and Pharmacopeia Commission was as important to manufacturers as it was to the FPA.

Compulsory dating remained unresolved in 1959. That year the BSI's new Director Professor David Glass dubbed dating in the draft standard currently in development as 'merely... a note expressing a pious hope' and not a declaration. This thrust the association into re-negotiations with manufacturers to ensure the BSI goal 'to make the date requirement mandatory'.[44] The BSI technical committee had a complete draft standard by 1955, however, the question of dating and the institution's commitment to review and if necessary revise the FPA condom/sheath testing in order to update the methods to meet modern rubber regulatory standards delayed its release. The association's rubber inflation, strength and ageing tests were scrutinised and modified in collaboration with BSI experts and manufacturers. Condom/sheath inflation test then involved attaching the item to a 'leak proof apparatus', filling it with air and then allowing it to stand for fifteen minutes before assessing deflation and airtightness. The strength test assessed the weight each product could endure before breaking. And finally, after heat-induced ageing, leakage and strength test were readministered to ensure the products maintained quality and efficiency over time.

In 1958, a water-based alternative inflation test was suggested: the bursting test. It worked by filling the item with three litres of water and wiping it dry to discern leakage.[45] At a 1958 technical committee meeting, Prentif's representative declared the company opposed all inflation testing. This came as quite a shock as Prentif had always keenly cooperated with the NBCA/FPA and willingly adapted its own early 'factory tests' that were early forerunners of the formalised inflation tests to meet association standards, as well as actively trading off the use of a 'guaranteed tested' label for advertising.[46] However, now that there was a chance these might be formalised and enforced, and in light of its decreasing market share and refusal to adapt its manufacturing processes to match the more efficient modern methods its competitors were using, the company announced that it had determined that 'both the proposed methods give rise, by virtue of the stresses they impose, to faults in an otherwise serviceable article'. It demanded a more technologically innovative method. This position was fully supported by London Rubber who in 1953 developed an electronic test that was applied to assess and maintain its own standards. This worked by placing each condom over a probe and sending a charge through it; if any charge was detected outside the condom a fault was indicated and the product scrapped.[47] Manufacturers claimed that this method alone, prevented rubber goods from the undue stresses of association approved testing, that imposed arbitrary and artificial pressures that were well outside those of normal use, and should thus 'eliminate both the [inflation and bursting] methods' from inclusion in the standard. In response, FPA testing expert George Carruthers, who had personally tested over '300 rubber condoms' using the methods in question, ignored the demand for a new test, and argued that both tests subjected products to 'adequate strain' and were 'equally satisfactory' for their purpose.[48] However, he suggested that the water test had one advantage, it could 'be conducted more rapidly and several tests conducted simultaneously'. The technical committee agreed the bursting test was 'simpler' and made it 'easier to detect any pinholes'.[49] Thus, the test that ultimately won out was the least technical and laborious and these two features were deciding factor. This demonstrates that the BSI standard, like all the FPA's testing work before it, was based not on innovation but effectively guaranteeing and standardising the products on the market. There was little overt ideological motivation or influence, just a keenness to meet its duty to share knowledge, innovation and best practice to ensure quality, efficacy and safety of all the products it accredited.

The last issue that seriously threatened the feasibility of the rubber contraceptive standard in the mid-1950s was cervical cap spring tensions and dome heights. Earlier these issues had proved contentious for the FPA's regulatory agenda and the limitations of standardisation combined with subjective nature of patient needs and doctors' preferences had resulted in the availability of a variety of products with diverse dimensions. But in the 1950s the BSI technical committee demanded that firm decisions and standards regarding caps be negotiated and set. Once again, the association was expected to take the lead in negotiations and determine and effect the best outcome.

Diaphragms and caps were the most complex rubber contraceptives to standardise. They were designed for long-term use, had to be applied internally which hampered the verification of perfect use, sizing was challenging as individual women required varied styles and sizes over their lifetime, and finally, in normal use, it was possible to fundamentally damage the product's structure by snapping the spring in the rim or tearing the rubber.[50] The FPA had confronted all these issues when setting approved list standards, but had faced barriers in fully understanding the needs and issues associated with the technology as many patients failed to return for follow up appointments; and those that did reported conceptions despite wearing the device, found them uncomfortable or ill-fitting, experienced difficulty in insertion or removal, or found them to be too rigid or flimsy, or, thick or delicate. Despite years of attempting to ascertain answers to these questions, the pure and applied scientific efforts of the association failed to completely address all the issues with caps. But this BSI collaboration provided it the power to compel acceptability, efficacy and manufacturing data to attempt to address these ongoing areas of concern in the forthcoming standard.[51]

The technical committee widely solicited medical and manufacturers' opinions on the most important aspects of diaphragms and caps and assessed the models and brands which the association and manufacturers' studies proved most satisfactory to patients and effective in use. Much of this data was of course coloured by the commercial and regulatory interdependence of the association and its suppliers and their collaborative efforts to refine and standardise the female barrier methods then in prescription. This streamlining meant that earlier determinations regarding the needs and preferences of the association and its physicians and patients had aligned with the manufacturing and business abilities and agendas of industrialists to somewhat limit the market, and thus, steer the evidence that was solicited by the BSI toward products, brands and styles that already formed the basis of the mid-century cap/diaphragm market. Nonetheless, the BSI used the surveys it commissioned to determine the minimum necessary contraceptive cap standard and identify any important regulatory elements the approved list had neglected.

In 1954 the FPA included ten varieties on its approved list; these represented a large range of styles and brands, from racial, 'prencaps', vimule and dutch caps made of rubber in up to five sizes with both flat and spiral rims, to plastic varieties and the flatter and wider latex diaphragms.[52] There was no advice included that spoke to the testing methods and the only specification relating to caps was that they must be prescribed and fitted by a physician. Even the association's standards proposal offered very little firm guidance. It stipulated that all diaphragms and caps must have their size marked on the actual product or its individual storage container, that measurement was to be based on the circumference of the outer rim for diaphragms and the inner rim for caps, and that a *maximum* ± 1 mm leeway was to be given for all sizing.[53] Four years later when the work to define rubber standards was much further advanced, a list of tests and guarantees was included that assured firm 'standards for durability, ageing and distensibility of rubber and plastic appliances' were now defined that could guarantee the quality, efficiency and harmlessness of these products.[54] In reality, this claim spoke more to the status of the condom/sheath standards than to the caps, for which requirements, norms and ideals remained almost as uncertain as they had been decades earlier. But, based on the results of its general rubber tests and standards, the satisfactory cap and spermicide prescription that had long been successfully applied in FPA clinics, and its strong reputation as a service provider and keen, but limited, regulator, the association took responsibility for these claims through the introduction of an 'Approved by the Family Planning Association' stamp of warranty.[55]

It was this gusto and commitment that the association took into consultations with the BSI to alert the institution to the issues it had experienced setting shape and height standards for comfortable and acceptable cap fitting. For manufacturers and contraceptive suppliers these were minor concerns as they already issued pamphlets and guides detailing the correct insertion and extraction of devices and offered personalised instruction and fittings provided by specially trained nurses and physicians.[56] But for regulators and guarantors, these matters remained extremely complicated. Even in the mid-1950s, the optimum structure of vaginal and cervical caps remained unknown. The lack of such basic knowledge prompted the BSI to instigate a medical survey to analyse and agree a minimum standard structure for each type of female rubber barrier method. Eleven doctors specialising in contraceptive medicine returned the BSI's diaphragm/cap questionnaire. The responses demonstrated that practical medical opinion was generally uniform. Ten of eleven doctors preferred 2.5 mm increments between available diaphragm sizes, but five out of ten preferred 5 mm size increments for caps; the remainder stated no preference. Further, most agreed that manufacturers should produce caps in three regular sizes, and any falling outside those parameters could be made to order.[57] All practical and easy solutions. After uniform sizes were set, the issue of dome height remained problematic to the establishment of a standard. Only Prentif manufactured high vaulted caps and the rest only low.[58] Medical opinion was evenly split; as explained, many physicians had a cap or brand preference, but most relied on patient acceptability to direct prescription. As there was a clear market demand for both high and low domes, the BSI, like the FPA had done, worried that standardising caps may be impossible. Prospects looked especially bleak when Prentif threatened to discontinue the high dome caps if the standard did not contain specific height requirements to guarantee an ongoing market for this product.[59] So, the need for both necessitated that no single dome height could be agreed to be *the* standard.

The final matter impeding the creation of the rubber standard was the appropriate spring tension for diaphragms and caps. Early in the 1950s the FPA tried to persuade manufacturers that the available spring tensions were only appropriate for small and medium diaphragms and caps, and that higher tensions were needed for large-sized caps. This request was an end-user requirement that resulted from the association surveying the different contraceptive needs of its patients; but making diverse springs was unfeasible from an 'economic point of view', and manufacturers opposed this suggestion.[60] London Rubber expressed the greatest opposition to dual spring tensions. It unilaterally declared it would offer only one kind. But in the interest of cooperation London Rubber did offer the association the choice of which tension was superior: regular or high. This was absolutely not in keeping with the association's goal to provide patients, especially those requiring larger caps, effective and safe contraceptives and although the FPA 'urged the manufacturer to retain the present tension' and threatened to actively prescribe patients away from the brand in future, this was near impossible.[61] As Borge has shown, by this time the association had grown dependent on London Rubber for the bulk of its merchandise and despite wielding significant authority and trust, could really do no more than threaten the manufacturer who merely desired legitimation and approval from the association, but by the 1950s no longer needed it.[62] This was a new world, filled with contraceptive, regulatory and purchasing prospects and the old battles for legitimacy were now being usurped by fights for market and social capital and brand recognition and trust in the increasingly competitive battlegrounds of advertising and consumption.

In 1956 the BSI technical committee on surgical rubber and plastic appliances admitted that the spring tension and dome height issues each necessitated substantial investigation to resolve.[63] Thus, two separate rubber standards were needed: one for condoms and washable sheaths and another for diaphragms and occlusive caps.[64] As the majority of the testing and language for each standard was identical both were considered jointly until 1963[65]; after which they diverged as the condom standard was considerably advanced. Although work continued apace at the FPA, all consultations regarding the cap standard remained 'informal' until after the condom standard was released.[66] In earlier chapters, I have evinced the tendency of officials and non-movement aligned activists and groups to delay and drag out contraception-related research and decision-making. This was also the case with both the Institute and the Pharmacopeia Commission. Each was aware that this work ultimately would fall to them and made necessary preparations towards that end, but was yet in no rush to effect that outcome. A key bellwether for grasping public feelings regarding contraceptive acceptability and keenness for quality and efficacy accountability came in the form of the 1963 Consumer's Association *Contraceptives: A Which? Supplement*. This product, method and standard survey essentially achieved everything the FPA had wished of its approved list but with far greater reach and imbued authority than the association, and its local and global counterparts including Marie Stopes International and the International Planned Parenthood Federation, had yet achieved through independent and collective medico-scientific, promotional and regulatory labours.[67] What advocates did achieve was the wide dissemination of new and effective contraceptive methods as they emerged and were refined. But ironically, competition and antagonism, largely presenting as endeavours to disparage and discredit competitors, between the very contraceptive manufacturers whose products and brands were advertised and actively promoted by the association, negatively impacted mass acceptance and trust of mechanical contraceptives.[68] It was this lack of trust and clarity that led to the Consumer's Association publication. It claimed that the assessment was conducted as a result of public demand in spite of there already being 'a means of [public] protection, though not many people seem to make use of it'—the FPA's approved list of contraceptives.[69] That is obviously true, but there were significant limitations to publicising the list that meant it never became a popular source of contraceptive surety: primarily, it had to be actively sought out by medical and lay consumers. This meant that potential readers had to be both accepting of and interested in having a surety associated with contraceptive products, as well as knowledge that the list existed and a willingness to engage with the association to obtain it. All of these demands required significant investment, engagement and dedication that even the association's own statistical studies acknowledged were shown to be difficult to obtain, and even harder to maintain. Even in the early 1960s, after the pill had been released for sale, there was still personal, professional and official reticence to engage with and rely on the FPA's science- and clinical evidence-based claims. This would soon change—but that was more due to the regulatory needs inherent to the pill, than a real association victory.

This in part explains why the British Standards Institute negotiations were dragged out and why the British Pharmacopoeia Commission had taken almost no steps to include chemical contraceptives by 1956. In fact, the monograph the FPA had submitted to the commission's technical committee was sidelined; despite the association completing extensive revisions of its efficacy and harmlessness tests to meet 'modern biological standards' and appeals to the Medical Research Council to legitimise its proposed methodology.[70] Finally, after a further three years of dragging its feet, the commission admitted that it 'had decided against including a formulation for a contraceptive in the B.P. because... the inclusion of any controversial material might antagonise... and prejudice the sales of other products'.[71] This was of course completely outrageous as, is discussed below, there was by 1959 a revolutionary biochemical compound being prescribed as a fertility therapy that had contraceptive side effects.[72] In light of this fact and its apparent, but slow, success having contraceptives covered by the British Standards Institute, the FPA refused to relinquish its goal for the Pharmacopeia Commission. That year the association planned to push the matter by having influential supporters petition the commission to cooperate with the FPA.[73] Unfortunately, that goal was scarpered by BSI's announcement of the condom standard's delay. This proved a particularly tough blow, as the standard then drafted proved near identical to the version that was finally published five years later.[74]

From mid-1960 the BSI technical committee repeatedly advised the association that the standard would 'be agreed [and announced] fairly soon'.[75] However, FPA attempts to confirm the release date were put off with repeated promises that the BSI 'hoped to issue a standard in 4–6 weeks'.[76] In truth, the Institute was having trouble convincing its sampling committee to approve the standard as composed; the proposed sampling method, to select a random number of contraceptives from each 'quantum of production', was rejected as too vague, and the inability to agree to an alternate method threw the entire publication into doubt.[77] Finally, on 8 April 1964, the efforts of the association and its friends at the BSI culminated in publication of 'the first in a proposed series of specifications for contraceptives', this one specifically covering condoms (B.S. 3704). It was concurrently announced that London Rubber the nation's largest condom manufacturer had received the first and only approved contraceptive Kite Mark, the BSI certification stamp, that advised consumers its products 'comply in all respect with B.S. 3704 and are BSI-Approved'.[78] Borge has shown that although London Rubber initially believed this to be an unnecessary incursion into its autonomy, it also provided the manufacturer the endorsement it had long been keen to obtain from the association; and from an even greater authority. This was a win–win for all involved, with the exception of other contemporary condom manufacturers, who were initially excluded from receiving a Kite Mark until they could update production methods to produce condoms commensurate with the new rubber standard.

Following the release of the condom standard, work resumed on the diaphragm and cap standard. The BSI committee ultimately determined that some matters, like spring tensions and dome heights, could not be fully accommodated and opted to instead focus on controllable matters such as the definition of size intervals, and the approval of two varieties of spring: helical and clock. It omitted dome shape or height and purposely ignored 'spring resistance' except to state they 'should provide suitable resistance to deformation'.[79] Two years after the first, the Standard for Diaphragms and Caps (B.S. 4028) was released; but it was into a world where female-oriented and controlled hormonal contraceptives dominated national debates and news cycles owing to their novelty, ease of use and asexuality.[80]

Chemical contraceptives were similarly dragged into the repertoire of the British Pharmacopoeia and its Codex in 1963 and 1968 respectively. But these were incorporated with far less fanfare and more stealth. This inclusion could be used to argue that a demonstrable social and cultural shift to scientifically and medically guarantee all contraceptives had taken place in the early 1960s, but in the case of the pharmacopeia, it must be noted that the reasons it had long cited for denying the inclusion of spermicides were fundamentally scientific; it determined it could not sufficiently define a chemical testing method because some common base ingredients could not be satisfactorily standardised and thus the spermicidal efficiency of the products could not be guaranteed. Further, it was not willing to certify the spermicidal formulations as even the manufacturers, specifically H. G. Rolfe of British Drug House, admitted that different processes of mixing and manufacture meant 'that the production of [contraceptive] preparations… is an art, not a science'.[81] The reality was that even when contraceptive methods and practices were based on and arbitrated using scientific principles, the human factor inherent at every stage of design, manufacture and use made regulation and standardisation a precarious venture. The FPA was able to undertake and pursue this work in the unofficial niche it had carved between illegitimacy and expertise, but persuading other regulators to trust and rely upon its scientific standards was a more challenging pursuit. The tipping point was reached due to a concurrent influx of social, political, medical and actual pressure brought to bear by the emergence of oral contraceptives. These raised legitimate and pressing questions about the association's authority and experience regulating modern contraceptive technologies that undoubtedly required formal medical oversight and prescription. The future of the association and its ability to continue to (try to) shape, provide, and regulate the British contraceptive market was contingent on its ability to adapt to and oversee a whole new technology in a rapidly changing medical and cultural marketplace.

The Medical Profession, Association, and Hormonal Contraceptive Science

The British medical profession's long-term antipathy and uneasiness toward even the principle of contraception, let alone its practice, has been well documented by historians. But this was gradually overcome and by the early 1960s with the increased acceptance and support for the various methods and practices of fertility control (with the exception of abortion) that followed the emergence of the first undoubtedly scientific birth control method: the synthetic hormone-based oral contraceptive pill. This transition was anticipated by Western family planning advocates; as since the 1930s research was undertaken to use hormones to induce temporary and/or permanent sterility. This work was inspired by the American medical profession's insistence that although existing contraceptives had 'proved quite satisfactory', a 'simpler and better method' was required to address population concerns.[82]

From the mid-1930s growing numbers of British doctors embraced birth control and endeavoured to study and apply it in accordance with the medico-scientific methodologies then taught by contraceptive clinic physicians, notably those affiliated with the NBCA/FPA. One of the earliest published expressions of concern came from Professor of Obstetrics and Gynaecology James Young, in the 1937 *Journal of Contraception*. He openly criticised the wider profession's apparent lack of interest in the topic and warned that unless its members faced the practice 'freely and frankly', physicians would inevitably find their influence over sexual, maternal and contraceptive health severely compromised.[83] Significantly, his warning was bounded by items examining the nascent studies of hormonal means of conception control, recent developments in endocrine physiology and promising animal experiments.[84] Research investigating sex hormones and their relationship to fertility and human reproductive cycles was widely available in British, European and American obstetrics, gynaecology and general medical presses and discrete and collaborative transnational studies instigated between scientists and family planning societies.[85] The profession was on notice that the burgeoning field of endocrine science would in future incorporate contraception.[86]

Local contraceptive advocates and scientists also viewed hormonal research and discoveries as momentous. The Birth Control Investigation Committee (BCIC) and Eugenics Society each expressed eagerness to instigate hormonal contraceptive research, however the NBCA/FPA was more pragmatic and insular in its mid-century focus and so downplayed this to focus on its standardisation and regulatory activities. Yet, despite this pivot away from potential hormonal research and development work, the association kept abreast of developments, but didn't fully engage until traditional contraceptives had been verified.

In *Disciplining Reproduction*, Adele Clarke explains that until 1925 the United Kingdom was the apex of reproductive scientific research and experimentation. British scientists combined local biological, medicinal and agricultural expertise to advance understanding of human and mammalian reproductive processes, especially the role of hormones in menstruation and their bearing on fertility and sterility at various cycle stages. British endocrinologist Francis H. A. Marshall, proved the ovary was the source of mammalian sex and pregnancy hormones and identified their reproductive role.[87] This discovery was complemented by Walter Heape, who pioneered hormone-based research into animal 'artificial insemination'.[88] These ground-breaking local studies informed contemporary knowledge, but, after 1926 the global epicentre of hormonal research shifted to the United States heralding the beginning of the 'heroic age of reproductive endocrinology'. This migration resulted in a fuller understanding of hormonal sex, growth and fertility systems, and ultimately prompted scientific advances towards the development of oral contraceptives.[89] Eugenics society, BCIC and association-affiliated and funded scientists significantly contributed to the 1930s 'endocrinological gold rush', advancing understandings of the hormonal control of sterility and fertility in animals and humans.[90]

In 1934, association researcher Solly Zuckerman teamed up with Alan S. Parkes of the National Institute for Medical Research, a Medical Research Council subsidiary, to investigate observations that 'animals fed [or injected] with [certain] hormones remained sterile while the doses continued'.[91] They further proposed to study the relationship between hormones, the menstrual cycle and 'fertility period' of the baboon, as the female demonstrated 'no essential physiological or morphological difference' with humans. The Eugenics Society and BCIC funded the enquiry as the researcher's submitted proposal strongly inferred practical uses for the contraceptive agenda: the possible development of a hormonal contraceptive, and/or a definitive understanding of menstrual cycle fertility fluctuations to scientifically support the 'safe period' which was then 'the only feasible contraceptive method for Roman Catholics'.[92]

This hormonal research could address the separate and shared contraceptive goals of the Eugenics Society and BCIC, and both agreed to extend funding as the potential human application of a compound to induce temporary or permanent sterility, presented shared, but ideologically different, population and fertility control opportunities that would render all other, less certain methods obsolete.

Zuckerman and Parkes' menstruation research discovered that oestrogen and progesterone levels varied at different times in the ovulatory cycle and that uterine bleeding followed the withdrawal of either or both hormones.[93] The function and timing of sex hormones within the ovulatory cycle were thus determined through this Eugenics Society and BCIC funded research, and borne out in Zuckerman's subsequent monkey and human studies.[94] These proved that daily application of oestrogen, prompted endometrium build-up, and shedding when oestrogen was discontinued. However, the findings were hard to duplicate in 'intact monkeys'. If this were the sole hormone involved, conception control would be hypothetically feasible by checking natural oestrogen functions. But contemporary experimental biologist also had to factor in progesterone which had in separate monkey trials delayed menstruation—in minuscule doses. There was a palpable sense that this transient, cyclical hormone was key to understanding and controlling menstruation.[95] These mammalian discoveries were intended to aid understanding of human reproductive and endocrine systems and when human testing was possible, evidence demonstrated these functions correlated closely in both species.

By 1950 the vanguard of hormonal inquiry had shifted to the bedside and treatments for menstrual and fertility conditions—now understood to be endocrinal—belonged solely within the medical domain.[96] Hormonal 'conception control' was theoretically no exception, but was not yet of 'paramount importance' despite being projected to become so shortly.[97] Owing to this indeterminate and still largely theoretical forecast the NBCA/FPA was reserved about discussing or endorsing any possible hormonal contraception.[98] As a result of this caginess and failure to publicise and promote emerging hormonal contraceptive tech, the local medical profession was able to maintain its stance that all contraception was essentially a non-medical practice, and limit professional engagement with it into the 1950s.

Nonetheless, between 1930 and 1950, hormonal therapies for sexual and fertility failures were being sporadically tested.[99] In the late 1930s association and Eugenics Society affiliate Carlos Blacker advised other physicians treating patients 'apparently suffering from "temporary" impotence' or erectile dysfunction to inject them with '5mgs of testosterone propionate' as a reinvigoration therapy.[100] Contemporaneously, hormones were considered as a possible remedy for male homosexuality. Blacker asked Zuckerman about this possibility and it was agreed that though they were 'doubtful about the possible efficacy of testosterone', and suggested it might actually strengthen the patient's sexual impulses, if no options were available, a trial 'may be of some use'.[101] It is a bit odd that the local physicians involved with developing, testing and trialling contraceptives with and for the FPA would be so risk adverse regarding hormones, when others amongst the ranks embraced the laissez-faire nature of scientific and medical testing and therapies in mid-century Britain, where drugs could be marketed without any oversight and practitioners had the ultimate discretion prescribing therapies, even experimental hormonal ones.[102] Novel approaches to sex problems and fertility and sterility human trials were eagerly reported and debated in the British medical press demonstrating the professionalisation of sex research and therapies had been achieved and integrated. However, this tolerance and engagement render the profession's apparent surprise at the innovation of hormonal contraception, and thus ongoing opposition to contraception's medico-scientific grounding, confusing. Even interwar hormonal treatments and therapies received collective medical focus, and so the elaboration of endocrine-based solutions for sexual and reproductive issues must surely have been anticipated; especially as the leading national contraceptive provider, all the major British manufacturers and indeed the global planned parenthood community and its representative body, the International Planned Parenthood Federation, were actively and loudly promoting the field's medico-scientific pivot and achievements. Ironically, just as earlier American-funded contraceptive enquiry was centred in Britain to avoid overt political constraints, the then unanticipated British medico-scientific inertia regarding practical endocrinology led North America to become the post-1935 epicentre of hormone-focused contraceptive research. Subsequently, the matter almost disappeared from NBCA/FPA attention until mid-century when rapid changes began to alter the local medical and contraceptive landscapes.

Collectively, but obviously not universally, physicians, mainly members of the British Medical Association (BMA), avoided the contraceptive debate prior to 1950. This was in spite of the two major 1930s concessions supporting the practice: the Ministry of Health's sequential concessions for medically necessary prescription at government health clinics, and the therapeutic recommendations of the Departmental Committee on Maternal Mortality and Morbidity. The profession's representative body also practically ignored the association's medical teaching work, its approved list of contraceptives, and rigorous Therapeutic Substances Act (1925) inspired and accordant testing and regulatory practises.[103] This Act made local Medical Officers of Health arbiters of drugs included in the *British Pharmacopoeia* and *British Pharmaceutical Codex* and provided them authority to verify the purity and safety of any given drug. So, if the FPA could have convinced the pharmacopeia commission to include chemical contraceptives, a formal integration of the divergent branches of contraception, medicine, science, and drug/chemical oversight would have been in place before the advent of the pill.[104] But this was not to be, so the association's medical reach and prestige were limited to influencing individual practitioners by imploring each to work within their conscience and experience on an as-needs individual patient basis. Even in the face of these limitations, the NBCA/FPA appealed for interested practitioners to assist in designing and directing contraceptive programs, but the BMA remained adamant that as long as it, seemingly arbitrarily, determined and maintained the public undertook the practice primarily for social reasons, it was not its duty to provide.[105]

Undeterred and knowing that its cause would eventually prevail, the association plodded along until 1944 when an opportunity emerged for it to cooperate with the BMA, who now conceded that there were some definitely medical circumstances requiring guaranteed contraceptives. That was the year the Royal Commission on Population was convened, and the association rightly recognised that the commission's interest in its work would garner positive publicity for its contraceptive and regulatory practices. So, the FPA invited the medical association to appoint a representative to its medical sub-committee. Considering the appeal, the BMA agreed that 'from the names of the medical men associated with it' and its expansion into undoubtedly medical fields like sub-fertility and gynaecology, the FPA was 'evidently an entirely reputable body' but its activities were controversial. The medical association had to decide whether it should appoint a member to oversee and advise FPA medical techniques or 'should remain aloof'. In a bid to avoid controversy and placate its Roman Catholic counterparts, the ethical committee weighed in to advise against 'associating the BMA with any body of people advocating contraception', even one whose medical directors were renowned practitioners, and products and methods were scientifically standardised and guaranteed.[106] In a not entirely unanimous decision, the latter choice was ratified and with that the profession squandered this significant opportunity to influence the FPA's medical standards and practices and the formulation and direction of contraceptive regulation.[107] In order to reach this counterintuitive decision, the BMA alleged that the association's otherwise 'entirely respectable' physicians were actually employing an unsatisfactorily lax interpretation of medical 'need' to mass prescribe contraceptives; it effectively asserted that FPA doctors were working outside the remit and standards of the BMA.

Oddly and in direct contradiction with its own resolution, the BMA then attempted to dictate the perametres of the FPA's medical services and studies. It admonished that fertility work belonged in a 'fully equipped hospital department' and voluntary bodies should only provide such services from comprehensive medical sites, if and when local hospitals could not. It further ordered FPA specialists to maintain 'proper professional relationship[s]' with 'doctors responsible for the medical care of [clinic] patients', heavily insinuating a hierarchical relationship between legitimate areas of medicine and the lowly and unsanctioned FPA work. The BMA further advised that FPA clinics should run like hospital outpatient departments, where GP referral preceded treatment and a full report was returned to the referring doctor advising of treatment or prescription. This was an ironic proposition considering the BMA consistently rejected the responsibility to treat 'social' contraception patients.

Although the BMA response reads like a diktat for FPA clinical standards and practices, it had no real authority to make or enforce demands. Rather the association sought to build a cooperative relationship with the medical authority, similar to that it had built and re/negotiated with contraceptive manufacturers; it hoped to engender a mutually beneficial relationship where it could obtain some legitimisation and support from the BMA to undertake medical work it deemed outside its own remit, within accepted clinical and medical standards. However, this was not a shared aspiration of the medical association which preferred to alienate itself from the FPA and therefore medicalised contraception.

Ironically, this rejection occurred just as concrete hormonal contraception advances first emerged, and the Ministry of Health cajoled the profession to submit to a centralised and regulated health system. The National Health Service (NHS) Act of 1948, guaranteed comprehensive subsidised medical and dental services to 'every man, woman and child' in Britain, and promised doctors, dentists and specialists 'complete freedom to work for the Service or not'.[108] The Act promised to address 'shortages [which necessitated] new forms of organisation' and rescue public health from 'the maze, the unwieldiness, the overlap, the uneconomy, the lack of integration of [the extant] health system'.[109] The NHS would be directed and managed by the Ministry of Health. It proposed to streamline currently existing and overlapping services into fourteen districts to be managed by regional authorities through three core branches: (1) the hospital and specialist services; (2) local government services; and (3) various family practitioner services.[110]

The BMA initially opposed the NHS scheme. Ironically, in considering its repeated rejections of contraception on similar grounds, it alleged the proposal entirely focused on sickness and all but ignored the principles of preventive health.[111] It rightly asserted that his posed important questions about the physician's autonomy to practice within the system—but seemed blind to the fact that such concerns were also relevant to doctor's freedom to learn and prescribe contraception.[112] There was a notable blind spot for the BMA regarding birth control; it didn't, and couldn't, stop individual practitioners from engaging, or the association from attempting to medicalise and professionalise its services, but rather assumed most members would follow the body's leadership on the issue. But as Caroline Rusterholz and I have shown, this was not the case and doctors frequently went out of their way to learn from and later become accredited to prescribe contraception through the FPA.[113] In

Pressure Group Politics Harry Eckstein explains that the BMA generally supported the 'root and branch scheme for socialised medicine'; and similarly anticipated most association members would support its opposition to the proposed NHS organisation. However, when surveyed most doctors expressed support for the 'controversial' scheme, so the BMA 'disavowed the reliability of [its own] poll', and appointed a 'Negotiating Committee on the National Health Service' to consult with the Ministry on the profession's behalf. 'The power to speak [on important emerging issues] was thus gradually transferred from the membership as such, to a body in which older, established general practitioners were more heavily represented than any other group'.[114] This was directly comparable to the BMA's handling of contraception. The medical association unfailingly projected one, senior, stale perspective on behalf of the entire profession on an important matter of cultural and practical change. Meanwhile, BMA members increasingly flouted this authority to learn, teach and prescribe contraception and support the NHS reorganisation.

The FPA worked to engage the medical profession with its work until the new Act was introduced. This centralisation caused the association to pause and refocus its regulatory ambitions and petitions to the BMA and Ministry of Health. In 1948 FPA Secretary Margaret Pyke enquired with the Ministry regarding the Act's relevance to contraception and association clinics. She met with Senior Medical Officer in Charge of Maternity and Child Welfare D. M. Taylor, who clarified that through a 'combined voluntary and municipal effort' 'local authorities would finance the FPA to run clinics'.[115] Under this arrangement, the FPA would lose its autonomy and control and become subordinate to the Ministry and its direction and will. The association executive baulked and stated that under such an arrangement it 'had no wish, at this stage to be incorporated into the National Health Service'.[116] Thus, the BMA and FPA leadership were faced with the same prospect and each rejected what was rightly perceived to be a loss of autonomy to the state; but the association was able to resist the proposal and turn its focus inwards, this option was not available to the medical association.

Mid-century after the Royal Commission and NHS introduction, there was a discernible surge in public, political and professional support for the FPA. Despite years of efforts, the association had been unable to convince the press to cover its sites and their good works. All this changed in 1955 when a supportive new Minister of Health Iain Macleod offered to visit the flagship North Kensington site to demonstrate publicly his support for the practice and the organisation.[117] The visit was linked to the association's Silver Jubilee celebrations which provided both a legitimacy for the visit, and offered the still reserved press a 'family-friendly' angle to present the organisation and its services.[118] This wave of publicity and acceptability prompted the FPA to confidently apply for newly allocated financial support from the NHS that was to be made available to voluntary health societies to support geographical, physical and service expansion. All this led to the association becoming the central powerhouse overseeing and dispensing contraception and providing medico-social educational and therapeutic services to address marital, sexual and fertility difficulties. By the end of the decade the FPA had gained endorsement from the Ministry of Health, medical professionals and the public and was making definite progress toward its 'long term policy' goals of passing contraceptive regulation on to official bodies.[119] However, the hormonal contraceptive revolution reached Britain first.

During his April 1958 visit Gregory Pincus, the 'father of the pill', announced the discovery of a synthetic hormone with contraceptive properties to a meeting of 200 British specialists. His ground-breaking lecture was chaired by former President of the Royal College of Physicians and current FPA President Lord Russell Brain. His presence signalled both the drug's local and global contraceptive importance as well as its inherent medical nature. The British medical press admired the man, the drug's apparent certain efficacy, and the speedy institution of a 'field trial of [the pill] as an oral contraceptive'.[120] The *British Medical Journal* gushed that with Pincus working on hormonal contraception, 'it may not be … long before … a cheap contraceptive tablet [is discovered] which will be effective when taken perhaps only once a month'.[121]

The oral contraceptive, also known as the pill, was produced during two decades of intense steroidal and hormonal research and development in Europe and the Americas. Several motivated and charismatic individuals dedicated their careers, time and money to create the most revolutionary pharmaceutical ever conceived: a daily pill to prevent women from the 'threat [of near constant] pregnancy'.[122] In 1951 Carl Djerassi and Luis Miramontes working for Mexico City-based Syntex developed norethisterone, the first synthetic progesterone compound intended for oral consumption.[123] Other versions, notably G. D. Searle's norethynodrel, were quickly announced.[124] Through the financial and political support of American birth control advocate Margaret Sanger and longtime Planned Parenthood supporter Katharine McCormick, significant progress was made to create hormonal contraceptives in the 1950s. The pair funded Pincus at the Worchester Foundation for Biomedical Research in Massachusetts to develop a hormonal contraceptive. With this support Pincus tested norethynodrel's effects on animals, and later humans through fertility trials undertaken by gynaecologist John Rock in America and contraceptive trials led by Pincus and Min-chue Chang in Puerto Rico.[125]

Intense scientific research and testing into hormones, led Pincus' pill, Enovid, to be marketed in Britain and America before 1960. In the USA it was severely restricted pending further trials, but in Britain looser pharmaceutical constraints allowed the relatively untested product to be prescribed by physicians for specific fertility and gynaecological issues.[126] The drug was unleashed on the public under the sole control of the medical profession that had until then actively attempted to entirely abstain from contraceptive prescription. Admittedly, initially it was not framed as a contraceptive product, but rather a method to kickstart an unpredictable menstrual cycle, but this was short-lived and despite the enthusiasm of the medical press for the drug when Pincus announced it, the profession remained unprepared for the monumental changes it would mean to medical responsibility and practice. Medically managed contraceptives were previously limited to intra-uterine devices and caps both of which were sufficiently unreliable and unappealing and so could be successfully ignored by doctors who so desired. This meant that the FPA had, by necessity, been central to the dominance and direction of these practices, and that it was well placed through its medical leadership and 340 clinics by 1960,[127] to monopolise future medical contraceptive work.[128] However, even it was unprepared for the therapeutic and regulatory oversight and transitions demanded by oral contraceptives and spent much of the early years of the pill shaping the medical and scientific landscape for the drug's provision.[129]

Oral contraceptives prompted many medical and moral debates within the British medical community. Some doctors reflected on their seeming conflicting roles as healers, family planning educators and later, gatekeepers of the only certain, but still experimental, contraceptive[130]; but also their now prime role in political, social and individual perceptions of contraception. Some, already primed through their engagement and accreditation from the FPA, embraced the changing landscape of modern medicine where patients could ask doctors to prescribe specific contraceptive drugs and therapies,[131] and others, typically older, more entrenched therapists, yearned to remain a pillar of medical knowledge, integrity and advice distinct from the expanding pharmaceutical contraceptive industry.[132]

But between 1946 and 1959 British physicians, although wary, increasingly engaged with contraceptives. Doctors recounted issues with prescription and insertion of intra-uterine devices, diaphragms and cervical caps/pessaries,[133] and continued to muse on the potential for contraceptives to produce physiological and psychological changes in females; disturbing the cervix and uterus, causing infertility, loss of libido and inability to orgasm.[134] The safety and reliability of insertable contraceptives, specifically the 'Silver' or 'Gäfenberg' rings and Gold Pin Pessaries, 'lead authorities to condemn' these products as abortifacients not contraceptives[135]; and the 'Dutch Cap' or diaphragm proved difficult to correctly fit and hold in place, and was dubbed the cause of pelvic endometriosis in some users.[136] This designation gave focus and direction to doctors' concerns regarding pelvic infections and pregnancies, and as Rusterholz has shown this was a primary reason intra-uterine devices temporarily lost favour at the exact time a seemingly less invasive contraceptive became available.[137]

For the entire century doctors maintained that some popular female-based methods were unsafe, caused complications and occasionally severe health issues.[138] Post-war when use and demand increased, professional interest was progressively drawn to apparent dangers; this was reflected in the FPA's concurrent pure and applied investigations into links between its products and possible harms arising from intra-uterine device insertion, chemical irritants and fallible and frustrating diaphragms/caps. Many of these emerging and lasting concerns were primarily perpetuated by the medical profession and regulators successful evasion of contraception, despite the FPA's ongoing labour and agitation. This professional assertion of apprehension regarding contraceptive standards marked its unofficial entry into this regulatory field. Just as the BSI determined it had a duty to oversee rubber and the Pharmacopeia Commission was ultimately convinced that spermicides warranted standardisation and approval, the local medical profession finally accepted that somebody must oversee contraception; even if it remained largely sceptical that it should be the association who had a socio-political agenda in relation to the matter, or doctors, who at this point had very little independent expertise or experience.

It was into this hostile and suspicious environment that the oral contraceptive pill was introduced to market in 1957.[139] The FPA had, as I have shown, spent the last few decades focusing its efforts on building evidence of and support for the safety and efficacy of what were now passè and clunky mechanical and spermicidal contraceptives, and working relatively unsuccessfully to engender formal checks and balances on these products. Ultimately, the association's primary impact in shaping medical standards of practice regarding contraceptive provision came in its efforts to delay and shape the British role of the drug, on the grounds that there was a lack of official regulatory measures for the medical practice of contraception which were now necessitated by the pill. Citing its long-term dedication to scientifically guaranteed safety and efficacy standards, the FPA was able to effect and direct large-scale testing of the drug before approving it for sale in its hundreds of national clinics. This was achieved through its co-founding and sponsoring of a group to direct and oversee clinical trials to assess the safety, acceptability and efficiency of contraceptives: the Council for the Investigation of Fertility Control. The group became central to shaping and enforcing British contraceptive regulation.

As has been very well documented by many past and contemporary historians, oral contraceptives fundamentally altered the relationship between the medical profession and contraception. Following its launch, patients were empowered to approach physicians for the drug and this, at least initially, somewhat diminished doctor's medical role in the transaction to that of the prescriber. In many ways, the profession's influence over contraceptive guidance was limited and the FPA had established renown and trust as an authority in the field. This fulfilled the 1933 prediction that the medical profession's disregard for contraception would be to its own detriment when it became recognised as medico-scientific and doctors were perceived to be unqualified to provide it. A fast adjustment period ensued wherein physicians quickly sought and gained contraceptive proficiency to meet the public perception that doctors, as the gatekeepers of the pill, were solely responsible for ensuring the contraceptive products they prescribed and inserted were safe and effective.[140] Oral contraceptives were even more removed from sex than mechanical devices in addition to being non-medically essential 'lifestyle drug[s]', but the mechanisms of the pharmaceutical demanded doctors take potential health risks seriously.[141] This became a shared objective of the profession and association and marked the formal unification of their contraceptive agendas into the 1960s.

Making the Pill Safe for Family Planning Association and British Prescription

There has long been a legitimate fear associated with balancing the health and utility demands and needs of newly developed and marketed drugs with their potential long-term effects. In relation to the pill, the FPA and medical profession were each openly concerned and critical of the short time lapse between the synthesis of oestrogen and progesterone and their mass promotion and consumption; particularly as each might be deemed liable for prescribing the medication in the event that severe adverse effects emerged.[142] By 1960 the FPA had invested over thirty years in contraceptive regulation and provision, and risked its expert reputation if high safety and efficacy standards were not upheld for the pill.[143] In response to the pill's significant risks and challenges, the association and its medical allies instituted research projects and clinical trials to protect their influence as well as public health.

In 1957 a £30,000 grant from Captain Oliver Bird enabled the FPA to form the Oliver Bird Trust.[144] The Trust's primary goal was to 'establish an organisation for the proper clinical assessment of the acceptability and effectiveness of contraceptives, for which at that time there were no facilities'.[145] Its members, which included many FPA executive and medical staff, affiliates and alumni including Margaret Pyke, Margaret Jackson, Lord Brain and Helena Wright's practitioner son, Beric, founded the CIFC and appointed prominent gynaecologist Eleanor Mears as its medical director with an instruction to discover and examine the potential risks of hormonal contraceptives and redress extant legislative laxity regarding the drug.[146] Mears, the chief local pill expert and advocate, designed, ran and 'was closely involved with all the early trials of oral contraceptives in Britain', particularly those facilitated by the FPA.[147]

From 1957 global pill clinical trials were conducted. Objectives ranged from assessing optimal contraceptive hormone dosage to measuring hormones relative to side effects, determining the brands patients found satisfactory, those that revived malfunctioning menstrual cycles, and whether users of various races, classes and education levels were equally protected and competent to follow the drug's 'exacting' daily routine.[148] Most of these early tests emulated the initial Puerto Rican trials; therein a selection of fertile mothers, took a pill 'nightly from day 5 to 24 of the cycle, counting the first day of bleeding as day 1', subsequent courses were initiated following a week-long break.[149] Participants were further required to keep a faithful account of their tablet intake, sexual contact, any side effects and their severity and longevity, and finally, bleeding and menstruation, including 'spotting'.[150] As many of the initial trials proved disastrous despite subjects reporting they had faithfully followed directions; the CIFC and association were eager to devise and apply objective means to scientifically and medically evaluate oral contraceptives to ensure their effectiveness and safety prior to the method being prescribed at FPA clinics. The Pincus method was judged unsuitable to provide the level of surety the FPA required as many early trial participants fell pregnant or were provided incorrect doses.[151] These findings compelled more independent and thorough product testing by the FPA and its new research arm to determine safety, efficacy and optimum dosage for contraceptive effect before approval.[152] The relationship that developed between the association and CIFC was definitively hierarchical, the former was supplanted in the relatively limited regulatory authority it had managed to effect by the better funded, more independent and authoritative other. However, it absolutely understood and accepted that a loss of its scientific investigatory autonomy and self-direction was a worthy cost for the formalisation and officialism of medico-scientific contraceptive research. All of its subsequent work in the field was spearheaded, managed and subordinate to that of the CIFC.

Susan Junod and Lara Marks have described that in light of these concerns, the FPA and CIFC agreed to prescribe the pill only 'after it had undergone [sufficient] scrutiny', and that this determination ultimately resulted in a significantly delayed uptake of the pill as a primary contraceptive in Britain. This was largely due to the FPA's long held monopoly on contraceptive prescription and instruction—that in this particular instance could not be sufficiently offset or challenged by the commercial contraceptive marketplace or the local medical community.[153] In adherence to its commitment to endorse and prescribe only proven contraceptives, the FPA initially limited local prescription until the drug was independently verified. This was particularly challenging for the association from the perspective that it was a contraceptive advocate as well as a provider. It had always striven to encourage clientele and provide desired and appropriate methods to effectively plan families, but it found itself somewhat unable to control the publicity surrounding the pill as its effectiveness was well documented and advertised by manufacturers, adherents and the local and international press; a further limitation to its authority was that the drug could technically be obtained from any physician found willing to prescribe it. Despite decades of work to have the medical profession rely on or at least ally with its regulatory agenda, the pill proved both an impetus, but also a risk to its methods and standards. But with the CIFC equally keen and galvanised to define and implement original tests, to establish to its satisfaction the drug's efficacy and safety, there was a thoroughly respectable branch of contraceptive science on which doctors could and would rely.[154] The CIFC's caution about oral contraceptives was akin to that of the association and its medical staff, this restraint was somewhat infective and mirrored the FPA's concerns about the possible health and sterility impacts of long-term chemical contraceptives; following on from that lead, the CIFC initiated animal trials to evaluate the 'contraceptive formula in rats and mice' for teratological, carcinogenic or hormonal impact before human trials would be considered.[155] This didn't, and in reality couldn't, take that long as the drug was available for non-contraceptive sale even before the testing methods were defined let alone initiated, so shortly, several small-scale human trials were instituted in collaboration with the FPA. These tested for pregnancy prevention, 'human toleration', inhibition of ovulation, and attempted to determine the 'optimum [minimum] dose' to achieve these ends without side effects.[156] As a direct result of earlier clinical trials at the North Kensington Women's Welfare Centre and other FPA clinics, trial participants were required to undergo 'precautionary screening for [cervical] cancer' as well as periodic smears.[157] Larger scale trials followed, and volunteers were needed to facilitate the investigation. But also, public expectations of contraceptive clinicians and the CIFC's examination activities and pace had to be managed. Thankfully, the popular press was supportive of both groups by the close of the 1950s and worked with them in their regulatory drive. In 1958, the *Sunday Pictorial* railed against the total lack of regulation or approval Enavid needed before it was marketed, dubbing it 'appalling that such an important scientific achievement – with its possible dangers through inexpert use – should be allowed to sneak into the country'.[158] The CIFC's promise to ensure pill safety was repeated front-page news and to that end, the paper broadcast the Council's urgent request for trial volunteers; this call prompted over five hundred women to enlist.[159] Over the next few years, FPA sites became increasingly central to the CIFC trial schedule and functions. Volunteers were directed to the North Kensington Marriage Welfare Centre, and many other clinics, to participate.[160] In March 1960 the first 'reasonably large scale' assessment of the drug's efficacy, acceptability and dosage began at the Birmingham clinic under 'carefully controlled conditions laid down and supervised' by the CIFC. The FPA made its clinics, staff and willing clientele available to ensure the success of the wider contraceptive legitimacy and regulatory goals associated with this new technology. It was no longer an unofficial, unacknowledged advocacy body, but a legitimate and authoritative medical and educational provider with close ties to the scientific and medical communities and the express approval of the Ministry of Health. It was in this, at least semi-official, capacity that it provided sites and means to clinically and sociologically assess the efficacy of 5mg and 2.5mg dose pills as the association and national ideal, as the 10mg high dose pills were observed to cause the unpleasant side effects seen in clinical trials.[161] The Birmingham low dose trial quickly proved the lowest available dose was insufficient and resulted in pregnancies as well as 'relatively frequent, [but] usually slight' and diminishing side effects; the 5mg pill proved a completely efficient contraceptive.[162] This FPA-based trial and its proposed ongoing patient observation agenda ultimately formed the basis of the CIFC's June 1961 endorsement of oral contraceptives for British prescription.[163]

The CIFC and FPA were cognisant that hormone dosage posed potential health and safety risks to local pill users and agreed to work together to ascertain the absolute lowest adequate dose, as this was also likely to be the best dose.[164] After 1962 the wider community of British doctors also increasingly strove to 'ascertain the smallest dose that would be an efficient contraceptive', following the lead of the FPA and its position. As women were increasingly eager to employ the pill as their primary contraceptive, discovering the optimum dosage and limiting side effects became a universal research focus.[165]

Increased appreciation of the sometimes-debilitating side effects of the pill, led to the unification of these investigations: the potential to mitigate negative effects through dosage. Some patients quit early trials due to severe side effects and local doctors adapted personalised prescriptive techniques to mitigate these.[166] Different women tolerated different doses better than others, required higher levels of one hormone or the other, and for many the combined pill was unbearable, necessitating a new variety of pill, the sequential 'regimen' which incorporated only oestrogen in the first half of the cycle, and combined progesterone for the last few days.[167] Whilst the association under the guidance of the CIFC assumed initial pill studies, it remained vital and necessary that medical and government authorities with proper legislative and scientific resources must control the drug. There was a significant impetus for formalising the kind of oversights the FPA had long demanded owing to both oral contraceptives and the detection of the teratological consequences of the ingestion of thalidomide by pregnant women. These compelled the Ministry of Health to form the Committee for the Safety of Drugs in 1963, 'to advise ministers, doctors and manufacturerss on drug marketing, clinical testing and adverse reactions'.[168] This remedy for the previously lax drug oversight came too late for the 2000 British babies born with severe defects due to the Distillers Company Ltd failure to undertake independent safety tests on the thalidomide preparation it had purchased from German manufacturer Chemie Grünenthal, and actively promoted and sold in the UK. Like the current claims about oral contraceptives' efficacy and safety, thalidomide was advertised as being so benign it could 'be given with complete safety to pregnant women and nursing mothers, without adverse effect on the mother or child'. These claims were unfortunately later proven to have been completely untested and unsupported by fact.[169] The FPA, CIFC and Committee for the Safety of Drugs refused to allow a similar public health risk from hormonal contraceptives, but oddly enough, despite the huge publicity and support this agenda would have drawn, none publicly linked the drugs and their potential risks to foetal development. However, behind the scenes, answers were sought from manufacturers about the links between progesterone use and developmental abnormalities, and independent data collected through close observation of unintended pregnancies amongst pill users.[170] This demonstrates that the highly cautious atmosphere surrounding the potential unintended consequences to offspring of drug consumption was being actively applied by the FPA and CIFC to their pill prescription and studies; but that to maintain the general sense of optimism then surrounding modern contraceptive technology, this research angle was downplayed pending and absolute findings of an effect.

From the earliest Puerto Rican trials, significant and definite pill induced side effects emerged causing concern and demanding close observation of pill patients. This reality left many physicians, especially those in FPA leadership positions, apprehensive to prescribe unnecessary medications with the potential to generate unforeseeable long-term side effects. Association-wide, physicians were particularly unenthusiastic and the organisation as a whole responded to these concerns by seeking and effecting practices and protocols that dispersed clinical responsibility for pill prescription and oversight by informing all pill patient's GP's about brand and dose, and the particulars of the duration and any side effects experienced.[171] In March 1960 the North Kensington Marriage Welfare Centre's medical committee openly declared it had 'no enthusiasm regarding the present oral pill'.[172] Its members cited British pharmaceutical regulation as the cause, and refused to prescribe it until the CIFC made an educated and impartial ruling. As a sweetener to counter such pushback, the 1961 declaration of the drug's safety and suitability for clinic prescription was accompanied by the announcement of a scheme for ongoing, long-term clinical trials to be administered through association clinics. The North Kensington medical committee agreed to 'co-operate with the FPA and refer patients who asked for the "pill" to the [central] trials' but remained dedicated to its tried-and-tested methods and worked to downplay local pill demand until its novelty wore off.[173]

One of the prime medical objections to oral contraceptives related to their interference with the internal functions of women's bodies. Without methodical testing and the availability of credible, conclusive data from long-term clinical trials, many independent and association-affiliated doctors were worried about physiologic, psychologic and reproductive risks to women and Britain. Medical journals were full of concerns asking: 'If Nature decides that science has invaded the very heart of her domain, what terrible penalties may she inflict upon the females of the species?' And 'how long will her tolerance continue?'[174] It was these very fears, shared by a significant proportion of the medical community, that prompted the association, as Britain's contraceptive regulator, to delay in formalising and approving the method in its approved list or practice until the mid-1960s, and to commence an ongoing and expansive approval testing agenda for each emerging variety of oral contraceptive.[175] By late 1963 there were five: Conovid, Conovid E, Anovlar, Ortho-Novin, and Volidan.[176] The FPA's long-held and practised standards remained key to its regulatory vision and the maintenance of its identity as an authority and arbiter for all available contraceptives, even in this brave new world where even its own most experienced contraceptive physicians were unenthused about the future of their field. It was to break down this stigma that the FPA executive increasingly recruited and paid new association clinics to undertake and report on large-scale clinical trials, that each assessed the CIFC approved oral contraceptives for slightly different outcomes and used discrete methods.[177] Although they were officially enthused and supportive, the CIFC, FPA and the medical and scientific community worked together to expose the as yet, 'hidden problems' that each agreed would inevitably be associated with oral contraception.[178] In the shadow of thalidomide, many doctors were enraged by early claims downplaying dangerous side effects, and openly criticised manufacturers and scientists working on the drugs and the perceived 'fervent religiosity that has developed in its use'. In direct response, doctors were asked to observe and publicly report emerging 'bad side effects and serious complications'.[179] A further matter of concern during the early pill years related to the terminology and mythology surrounding the drug and its nature and function. Oral contraceptives mimicked the 'natural state' of pregnancy and ovulation patterns and were additionally observed to alleviate the severity of unpleasant menstrual symptoms such as 'pre-menstrual tension, dysmenorrhoea, migraine, breast discomfort, nervousness, irritability, excessive menstrual loss, irregular menses, acne, spots, greasy hair, lack of libido and sleeplessness'. As many of these symptoms were also pill side effects, the drug was contextualised as having faithfully simulated 'coincidental symptoms women experienced in their everyday lives'.[180] This language and belief downplayed the synthetic nature of the pill to highlight its successful replication of human fertility cycles. The danger was that in attempting to make hormonal contraceptive technology appear less artificial, minor side effects were being treated as normal, despite the fact that pseudo-pregnancy and menstruation were wholly unnatural states.[181] Many doctors expressed concern about the potential dangers associated with this framing of the problem.[182] However, manufacturers and supporters regularly debunked medical calls for caution and study by arguing that oral contraceptives carried 'no greater risk than normal pregnancy'.[183]

When Enavid was first approved and marketed there was extensive medical interest in its immediately observable side effects such as 'nausea, vomiting, general malaise, dizziness, breast discomfort and headaches'; as well as other, less universal or immediate results such as breakthrough bleeding, spotting, menorrhagia, gastralgia and weight fluctuation.[184] The latter were frequently dismissed, even in early CIFC and FPA trials, as 'subjective symptoms influenced by many emotional factors' which were 'often present in normal women'.[185] But in light of the medical call to action, the FPA formed a CIFC Clinical Trials Committee to work in collaboration with its extant Medical and Scientific Advisory Council to undertake deeper, clinic-based investigations into the positive and negative experiences of association clients on the pill. The group worked closely with and took its investigatory lead and direction from the CIFC and the Committee on the Safety of Drugs over the course of the decade to define, shape and oversee its side effect studies.[186] These were, due to the urgency of the studies, less independent and more closely aligned with those of the broader medical community.

Investigation into minor side effects began during the pill development process. G. D. Serle and other early pill producers Schering, N. V. Organon and British Drug House were, as Jessica Borge, Claire Jones and I have shown, ethically and financially invested in manufacturing goods amenable to the masses, so, reducing avoidable or unpleasant side effects through product refinement was a focus.[187] In the early 1960s, independent and CIFC- and FPA-affiliated doctors and scientists enthusiastically undertook this task and discovered that altering oestrogen and progesterone doses yielded positive results in reducing minor, common side effects, but frustratingly, lower doses also drastically reduced contraceptive effect.[188] Eventually success was achieved in both areas by reverting to the FPA tradition of adapting prescriptions to individual patient's bodies and needs.

After 1962, medical and FPA attention turned to more sinister emerging side effects: blood clots, carcinoma and impaired hepatic function. The first sign that oral contraceptives might have long-term health risks arose within months of their approval for FPA sale, when the association revealed it would investigate long-term effects of pill use following a patient's death from pulmonary embolism.[189] That week the American Food and Drug Administration (FDA) issued a statement confirming six Enovid users had died from thrombophlebitis, and the Swedish Medical Board instructed doctors to prescribe with caution.[190]

In 1962, the North Kensington Clinic demonstrated its ongoing mistrust of the drug despite its approved list inclusion and authorisation for FPA prescription. In a heated exchange, the medical committee challenged the CIFC's results and transparency, alleging that 'doctors were given ... only favourable reports on the oral contraceptive and hardly any of the unfavourable ones'.[191] In a letter to the Chairman of the Medical Advisory Council and former FPA President Lord Brain, the committee demanded concrete 'information about the experimental data which influenced [the CIFC] doctors to release these drugs for use'. This was a schism in the contraceptive lobby's ranks, as the committee strongly insinuated the CIFC had ignored possible emerging 'unpredictable long-term effects' and stressed the need for the committee to be open about its investigative and reporting process to alleviate the very deep concerns and apprehensions of contraceptive physicians who were ultimately responsible for actual pill prescription and oversight.[192] Mears responded, in a supercilious tone, that it would be 'usual for a council of experts of this nature to give all the particulars and developments which led them to advise certain lines of action to members of the body which they are advising'. Nevertheless, after providing means for 'doctors who are really interested to read all the relevant medical literature', she offered to personally manage any 'particular enquiries [from the centre] which are not answered' therein. Mears would also liaise between the CIFC and North Kensington Medical Committee to ensure future clarity and trust.[193] This exchange embodied the Foucauldian idea of continual variations of power and knowledge, as the realities of the dispersed dynamics of British contraceptive research, approval, knowledge and use both transmitted and undermined the scientific authority of these institutions; distinct from and deeply intertwined with the power and knowledge relationships between other medical and government authorities and users.[194] As this exchange ensued whilst the first British thromboembolism cases emerged, the CIFC realised its initial dismissal of the FPA physician's concerns demanded urgent revision.

Early responses to the idea the pill caused blood coagulation were somewhat hysterical with reports of short-term users falling victim.[195] This prompted a flurry of research into the relationship, but initially no locally produced evidence was found supporting a link despite the pronouncements of American and Swedish health authorities.[196] This sparked further examinations,[197] that in 1964 reiterated the general position that thrombosis was not a common side effect of oral contraceptives.[198] Calm was brought to the debate with demonstrations proving that rates of thrombosis in oral contraceptive users were no higher than regular community rates.[199] The CIFC and FPA released complementary research and practice statistics that showed only 2.2 per thousand pill patients might experience such symptoms. In 1965 approximately 400,000 British women ingested the drug daily, only around 880 women were anticipated to experience the issue.[200] The safety of the drug far outweighed the risk by their separate and joint accounting and neither organisation was willing to lose the ideological and professional ground gained in the battle to normalise and legitimise contraception for risks that low. But the wider, less politicised medical community continued to be alarmed by the figure and some, possibly overcautious, doctors advocated decreasing prescriptions to mitigate potential dangers.[201]

Although thrombosis links were not sufficient to prevent prescriptions altogether, many FPA clinic medical committees and staff agreed a legitimate danger was likely to present, and cautioned restraint in prescribing to its, by then, tens of thousands of annual patients.[202] Many association clinics were generally slow to implement 'Pill Clinics' for this very reason. However, by the mid-1960s propaganda campaigns pushing the drug in spite of its potential thrombosis risk, led to increased local pill demand and consumption.[203] The North Kensington staff reported being consistently overwhelmed at dedicated 'Pill Clinics', owing to the association's introduction of strict rules about three, six, nine and twelve month follow-up appointments for all oral contraceptive patients. Pill seekers also consistently overran regular birth control clinic sessions. This hindered the experienced medical staff who still openly preferred to prescribe proven traditional technologies despite the pill's popularity. This impasse was remedied with the addition of more weekly dedicated pill sessions to meet the overwhelming demand.[204] By 1969, research 'strengthened the belief that oral contraceptives [were] a cause of venous thromboembolism and cerebral thrombosis' and it is now accepted that blood clots are a definite but rare side effect.[205] The FPA's caution and delay prescribing the method meant that many early patients avoided this side effect.

As I have noted, the FPA had long investigated the possibility of a relationship between contraceptives and cancer. The association during its earliest iteration in the 1930s was alerted to a possible relationship between spermicides and cancer and developed and implemented tests to assess harm and later a program of collecting cervical smears to detect and monitor carcinogenic changes over time. Immediately following its formation, the CIFC overtook the FPA's chemical contraception and associated carcinoma research work; and although it gave weight to the association's advice, experience and methodologies, like the BSI and British Pharmacopeia, it decided to self-define techniques and standards.[206] Not long after this, the FDA in approving the pill noted that as oestrogens had been shown to cause cancer, a 'small carcinogenic risk' was inherent to the drug.[207] As a result of this, both the CIFC and FPA worried about the potential carcinoma-inducing dangers of hormonal contraceptives. This was a key reason cited to support the introduction of a CIFC testing program and to publicly justify the FPA's initial delay in prescribing the pill.[208] In 1960 the FPA's Medical Advisory Council issued a statement to the *Lancet* regarding the 'risk of carcinogenesis'; it explained that cancer had developed in mammalian test subjects supplied continuous hormones and that 'oestrogens [were] known to cause abnormal[ities] ... in the mammary glands'. But early on even the FPA considered the human risk minimal, as pill patients were mandated a monthly oestrogen interruption to allow uterine shedding.[209]

Regardless, the FPA, CIFC, FDA and global scientists, physicians and manufacturers investigated the carcinogenic risk of the pill. The North Kensington medical staff were open with their concern about synthetic hormones and initiated a local pilot trial amongst pill patients to complement its decades-old cancer smear program. From late 1962 physicians collected 'cervical smears as routine procedure from all patients at North Kensington aged 30 years and over, and patients new to the Oral Contraceptive'.[210] In addition to the FPA mandated quarterly assessments, annual follow up smears were taken to directly compare and detect possible cancerous changes.[211] The trial was later ratified by the FPA executive committee and extended to other sites.[212]

In 1965 the association's charter recorded its objective 'to take gynaecological or genito-urinary specimens for diagnostic purposes'. This sanctioned the FPA to formally expand its smear services and introduce 'cancer prevention centres' to its prescriptive and educational work.[213] As Junod and Marks have demonstrated, the only viable response to the risk was to watch and wait as 'the induction period of all cancers in man is long (15–25 years) and therefore the effects of [pill] compounds ... [wouldn't] be seen for many years'.[214] The FPA's assertive application of the cervical smear and breast screening 'scheme to [make it] routine procedure' was as pragmatic and proactive an approach as was feasible in the early pill years.[215] So, although it lost some significant regulatory ground to the CIFC, the association maintained its medico-scientific work where it was most vital, in clinics.

In 1963 another significant success in the regulation of contraceptives was achieved. That year the British Committee on the Safety of Drugs, also called the Dunlop Committee after the British Pharmacopoeia Commission's first Chairman Sir Derrick Dunlop, accepted the responsibility to assess, approve and oversee all drugs available in Britain, even oral contraceptives. Keeping with the FPA's contraceptive regulatory methods, the Dunlop committee proposed to achieve drug regulation through cooperation with manufacturers, not government decree.[216] Despite this claim of cooperation, it promptly declared that all British pill manufacturers must assess carcinogenic risk through 'long-term toxicity tests' on two animal species. In 1972 the initial results of these trials were revealed: a carcinogenic result was produced in some species of rats and mice by long-term use of high dose pills, but it was concluded that 'this evidence [could] not be interpreted as constituting a carcinogenic hazard to women' if oral contraceptives were used as prescribed.[217] This hesitant and not-at-all conclusive statement failed to address concerns that intermittently reemerged throughout the century.[218]

The final danger that emerged and demanded CIFC and FPA investigation and oversight was the 1964 discovery of pill-related liver damage comparable with that experienced by anabolic steroid users.[219] Immediately, analogous findings were described amongst users of the various contraceptive compounds, each demonstrating hepatotoxicity early in use.[220] Within a year more tempered research emerged to claim that the complaint was exclusive to older users and that younger, long-term users showed hepatic function well within acceptable range.[221] Jointly, the FPA and CIFC hurried a proactive risk response. Patients with a history of 'venous, thrombotic episodes or liver disfunctions' were excluded from all trials and Liver Function Tests were added to the mandatory smear schedule for association pill patients.[222] In 1966, the Dunlop Committee reported that 'mestranol when administered to rats causes liver damage', but also that rats were prone to the complaint in addition to thrombosis and carcinoma. It called for 'long term studies [to] be carried out' and the CIFC took up the baton.[223] Although the FPA was concerned enough to factor this in to its practice and prescription methods, particularly for older patients, as the vast majority of patients seeking the drug were young, typically just married, it waited on external research results whilst cautiously prescribing and overseeing pill patients. None of the major side effects discovered really impacted the association's practice or handling of the drug because its initial response had been very slow, highly cautious and fundamentally informed by scientific and medical evidence at every stage. So, when serious effects were added to the already long list of minor side effects definitively connected to hormonal contraceptives, the FPA was able to fall back on its regulatory history and practice to retain patient trust and calm concerns.

Fears about the pill's potential to cause death and debility ramped up in the mid-to-late 1960s, alongside scare mongering media campaigns like the 'Pill Kills' and hysterical reports of thrombosis, cancer and liver function study results. These led to medico-scientific and government-led interventions to investigate and mitigate serious consequences of oral contraceptive use.[224] Large-scale transnational studies were commenced and locally the CIFC and FPA worked to review and replicate all manner of local and international trials and their safety and efficacy results. In 1964, the Australian Drug Evaluation Committee in cooperation with the Dunlop Committee launched a large scale and long-term study headed by Edgar Thompson, a vocal sceptic of pill side effects.[225] Upon the release of its findings in 1968; Thompson announced he had determined an explicit relationship between the pill and thrombosis, as well as other severe side effects relating to cancer and liver function. Despite this, both the British and Australian studies agreed 'health risks [were] outweighed by the benefits of the drug as an effective contraceptive'.[226] The FPA practices and regulations were built to factor in many opportunities to identify and prevent harm, should the pill be found to provoke it. Its pill clinic sessions were very carefully managed from both medical and administrative ends, and promptly reacted to identified concerns. When it became apparent that verbal pill instructions were wont to be misconstrued or misremembered, written instructions were provided.[227] When pill clinic appointments and oral contraceptive patients became so numerous that it began to impact bookings and clinic function, a suggestion that nurses might conduct three-month follow up sessions was quashed as entirely inappropriate from both ethical and regulatory perspectives. If more patients were presenting, then more sessions must be offered.[228] But ultimately, even the addition of extra services was insufficient, with wait lists of up to six weeks, and in response nurses were allowed to see follow-up patients, but had to have their prescriptive recommendations formally approved by the on-duty physician.[229] A similar response followed suggestions lay clinic staff could dispense the drugs as well as repeat prescriptions for up to six months.[230] The seemingly high levels of organisational and administrative oversight built into the practice were both the direct legacy of the FPA's long-fought battle to professionalise contraceptive practice and legitimise its practical and regulatory methodologies; and also a reflective response to medical and governmental actions to support the mitigation and treatment of serious pill side effects.[231]

In May 1967 the British medical press published both the Medical Research Council and the Royal College of General Practitioners reports on oral contraceptive risks. Each independently confirmed a causal relationship between the drug and thromboembolism.[232] The work resulted from a multiparty research project undertaken by the Dunlop Committee, FPA and the Ministry of Health, that retrospectively assessed sixty physician's patient data that was stored at the Royal College of General Practitioners Records Unit and Research Advisory Service.[233] This was in effect a more official and weighty version of the FPA's early case card-based sociological research, and was ultimately used to effect a great medical and public awareness of the efficacy, but also dangers associated with this wonder drug. It was the widest patient assessment yet, and aimed to expose the role of the pill in 'all deaths from thrombosis or embolism in all women of childbearing age'. Though the causal link was agreed, neither study could pinpoint the mechanism that caused blood clots.[234] Ultimately, after significant statistical research, lead investigator William Inman exposed the direct correlation between thrombosis and high dose oral contraceptives.

Before the medical profession could be notified, the discovery was leaked to the media. In mid-1969 the Dunlop Committee was forced to expeditiously distribute a 'yellow peril' flyer to doctors tardily confirming the danger of high dose oral contraceptives; this also prompted the government to 'warn doctors to no longer prescribe the higher dose (10-mg) pills'.[235] But the damage was done, and a media scare campaign was launched accusing the profession of complacency. This caused thousands of frightened and desperate patients to jam physicians' and FPA clinics and phone lines for placation and advice, and scared many into quitting the pill entirely.[236] The FPA was fortunate to be able to announce that in accordance with its own history of safety and effectiveness standardisation and contraceptive product and prescription regulation, it had many years earlier, established that low dose pills were preferable and had restricted all FPA prescriptions to 5mg and 2.5mg doses. It further pledged that 10mg doses would be entirely avoided in the future.[237]

Around 1970, women and doctors began to believe that they, like the unfortunate mothers who had taken Distaval a decade earlier, had been lied to about this pill's safety. By then global users topped fifteen million and the pill was by far Western women's first choice contraceptive. But anger began to supplant enthusiasm and activists, most prominently Barbara Seaman, criticised the drug. 'We were snookered', she explained. 'We were told that we were given this wonderful gift by modern science, that would make our lives so much better. But there was a very dark side... concealed and denied... And a lot of healthy young women died or were left crippled'.[238] In response the Dunlop Committee mandated that doctors must collect a thorough patient and family history (essentially a combination of the FPA's case cards and the Eugenic Society's family histories) and exercise extreme caution in prescribing the drug. It was further vital that clear and thorough information regarding pill risks reach users to ensure sufficient informed consent. So, a detailed booklet was included with every prescription dispensed.[239] At the close of 1971, 1.8 million British women were using some version of the pill, this figure represented 18 per cent of the entire fertile female population.[240] Although the definite association of the medication with serious, long-term and sometimes deadly side effects led a significant number of women to abandon the method, and return to older more hands-on but definitely safe methods, it also finally forced the Dunlop Committee, CIFC, FPA and medical profession to agree that formal contraceptive regulation was necessary.[241]

In 1970 eminent endocrinologist Alan Parkes edited the first edition of the *British Medical Bulletin* solely dedicated to the 'Control of Human Fertility'. He explained why the medical and scientific community had finally embraced contraceptive regulation. 'The answer is simple. The subject, with its ever-developing technology and its ever-expanding social implications, must be kept constantly under review, and, because of its vast importance, the more authoritative the review, the better'.[242] After half a century of FPA agitation and groundwork, the medical profession had officially accepted that it was the ultimate and correct arbiter of British contraception. Thereafter, the subsuming of the FPA's clinical work into the NHS was a logical progression and was not contested as it had been two decades earlier. In fact, early in the 1970s as its medical work and reputation remained well regarded, the association's medical leadership reflected proudly on their successes and anticipated the change. It had over its forty-year history founded the association and through it fundamentally medicalised the practice of birth control, defined and implemented scientifically viable standardisation and regulatory tools, and provided only guaranteed practice and products. This legacy helped the FPA navigate all the political and professional struggles it encountered and ultimately made it a trusted authority in the British contraceptive marketplace.

Conclusion

The emergence of effective hormonal contraception was a gamechanger for obtaining official recognition for and oversight of contraceptives in Britain during the 1960s. At the turn of the decade, the BSI and British Pharmacopeia Commission were both persuaded to work with the FPA to redefine and implement rubber and chemical contraceptive standards and approvals. The association was also instrumental in laying the medical and scientific groundwork that culturally aligned the practice of contraception in the perception of the British public, even before the medical profession would concede a duty in that arena. Despite its great and ongoing efforts to encourage and cajole the profession to oversee at least the more medical technologies such as caps, chemicals and IUD's, the profession and government refused to oversee and control contraception.

Only the pill, and really the emergence of seriously dangerous side effects, ultimately convinced official medical and pharmaceutical arbiters to accept the medical nature of the practice. The FPA, long aware of the potential risks inherent in the mass application of contraceptive technologies in general, let alone a completely new and novel method, had cautiously responded to hormonal contraceptives. It adhered to, and publicly highlighted that the reasons for its reticence were based on its desire to be properly informed by the science of the matter, not to uncritically accept manufacturer's safety and efficacy assertions as fact; but to independently prove each for itself. The association thus delayed the rollout of 'pill clinics' at its sites until its affiliates at the CIFC, had investigated and approved the method. As a result, when the first cases of serious side effects emerged, only a relatively small number of British pill patients were impacted. The association's restraint also meant its patients were prescribed low dose pills owing to the FPA/CIFC's early studies to determine the lowest safe and effective dose.

So, the FPA's slowly and painstakingly earned trust was maintained even as more official contraceptive oversight, by the CIFC, Dunlop Committee, Department of Health, and medical profession materialised. The formal opposition that had coloured earlier debates about standards and responsibility was slowly but surely eroded by the medico-scientific nature of hormonal contraceptives and the FPA's tireless efforts, on its own and in forming and maintaining networks and partnerships, to make contraceptive practice and its methods scientifically verifiable and medically palatable. Ultimately, its efforts helped to effect the mass surveillance of contraceptive technologies, the introduction of educational material to ensure patients informed consent and proficiency, as well as achieve a greater collective consciousness of established and emerging dangers associated with the pill.

Notes

1. Tanya Evans, 'Knowledge and Experience: From 1750 to the 1960s', in *The Routledge History of Sex and the Body, 1500 to the Present*, eds. Sarah Toulalan and Kate Fisher (United Kingdom: Routledge, 2013), 257.
2. Evans, 'Knowledge and Experience', 258.
3. Eustace Chesser, *Marriage and Freedom* (London: Rich and Cowan Medical Publications, 1946), 55.
4. Simon Szreter and Kate Fisher, *Sex Before the Sexual Revolution: Intimate Life in England 1918–1963* (Cambridge: Cambridge University Press, 2010), 327–9, 349.
5. Note by P. Cripps, 4 July 1952. Wellcome Library, Archives of the Family Planning Association (WL/SA/FPA), WL/SA/FPA/A7/23.2.
6. Letter R. Edwards to I. James, 18 October 1953. WL/SA/FPA/A7/23.2.
7. Response I. James to R. Edwards, 22 October 1953. WL/SA/FPA/A7/23.2.
8. Interview with R. Edwards, 10 November 1953. WL/SA/FPA/A7/23.2.
9. Letter J. Falk to I. James, 28 July 1953. WL/SA/FPA/A7/22.
10. FPA Internal Memo on Preparing a Monograph for British Pharmacopoeia or British Pharmacopoeia Commission, 27 July 1953. WL/SA/FPA/A7/22.
11. British Pharmacopoeia Commission Ad Hoc Committee on Spermicides Minutes, 15 September 1958. WL/SA/FPA/A7/22; Letter A. J. Clifford-Smith to M. Pyke, 16 September 1955. WL/SA/FPA/A7/23.2.
12. Submission to British Pharmacopoeia Requesting an Official Contraceptive Preparation, 5 October 1953. WL/SA/FPA/A7/22.
13. FPA Memorandum to the Royal Commission on Population, n.d. WL/SA/FPA/A18/1.
14. FPA Sub-Fertility Submission to the Royal Commission on Population, n.d. WL/SA/FPA/A18/1.
15. Letter FPA to C. Vian, 20 April 1945. WL/SA/FPA/A18/1.
16. FPA Evidence to the Royal Commission on Population Report, 27 April 1945. WL/SA/FPA/A18/1.
17. René Vallery-Radot, *The Life of Pasteur* (London: Constable and Co., 1911), 9–22.

18. FPA Evidence to the Royal Commission on Population Report, 27 April 1945. WL/SA/FPA/A18/1.
19. *Advertisements for Conception Day Indicator*, March 1957. WL/SA/FPA/A7/25.
20. Letter P. McCallum to R. C. Eleston, 24 November 1960. WL/SA/FPA/A7/25.
21. Letter A. J. Clifford-Smith to A. Boini, 25 April 1960. WL/SA/FPA/A7/25.
22. Letter G. B. Carruthers to H. J. Stern, 25 January 1955. WL/SA/FPA/A7/32; Letter A. J. Clifford-Smith to M. Pyke, 16 September 1955. WL/SA/FPA/A7/23/2.
23. Letter FPA Medical Sub-Committee to G. W. S. Blair, 28 July 1953. WL/SA/FPA/A7/22; FPA Internal Memo on Preparing a Monograph for British Pharmacopoeia or British Pharmacopoeia Commission, 27 July 1953. WL/SA/FPA/A7/22.
24. FPA and H. A. R. Binney Meeting Notes, 10 January 1955. WL/SA/FPA/A7/23.2.
25. H. M. Glass and FPA General Secretary Meeting Notes, 23 August 1955. WL/SA/FPA/A7/23.2.
26. FPA Medical Sub-Committee Minutes Extracts, 25 March 1947. WL/SA/FPA/A7/23.1.
27. Interview with R. Edwards, 10 November 1953, 22 October 1953. WL/SA/FPA/A7/23.2.
28. Letter R. Edwards to A. J. Clifford-Smith, 25 October 1955. WL/SA/FPA/A7/23.2.
29. BSI and FPA Rubber Standard Informal Meeting Report, 15 September 1955. WL/SA/FPA/A7/23.2.
30. Tests for Contraceptives on the Approved List, 25 October 1954. WL/SA/FPA/A7/23.1. BS 903 was the Standard covering vulcanised rubber goods.
31. Tests for Contraceptives on the Approved List, 25 October 1954. WL/SA/FPA/A7/23.1.
32. Letter I. James to A. R. Reid, 26 January 1954. WL/SA/FPA/A7/23.1.
33. Letter A. R. Reid to I. James, 19 January 1954. WL/SA/FPA/A7/23.1.
34. FPA Medical Sub-Committee and London Rubber Discussion, 25 March 1947. WL/SA/FPA/A7/23.1.
35. Letter A. R. Reid to I. James, 30 December 1953. WL/SA/FPA/A7/23.1.
36. Jessica Borge, *Protective Practices: The London Rubber Company and the Condom Business* (Montreal and Kingston: McGill-Queen's University Press, 2020), 107–8.
37. FPA Medical Sub-Committee and London Rubber Discussion, 25 March 1947. WL/SA/FPA/A7/23.1.

38. Borge, *Protective Practices*, 126.
39. Memo G. B. Carruthers to I. James, 30 September 1954. WL/SA/FPA/A7/23.1.
40. Letter I. James to P. M. C. Watkins, 14 October 1954. WL/SA/FPA/A7/23.1.
41. FPA East Midlands and East Federation Council Minutes, 27 June 1963. WL/SA/FPA/A7/23.2; Draft British Standard for Rubber Condoms, Washable Sheaths and Diaphragms, December 1959. WL/SA/FPA/A7/23.2.
42. Letter Prentif to FPA, 11 February 1954. WL/SA/FPA/A7/23.1.
43. Borge, *Protective Practices*, 76–8, 81–7; Claire L. Jones, *The Business of Birth Control* (Manchester: Manchester University Press, 2020), 173–81, 191–7.
44. Letter H. M. Glass to M. Jackson, 8 January 1959. WL/SA/FPA/A7/23.2.
45. Draft British Standard for Rubber Condoms, Washable Sheaths and Diaphragms, December 1959. WL/SA/FPA/A7/23.2.
46. Prentif, *Contraceptive Practice: A Quarterly Bulletin Dealing in Family Spacing and Allied Subjects*, 2 (February 1938). WL/SA/FPA/A7/23.2.
47. Borge, *Protective Practices*, 125.
48. 150 tests were undertaken using both the inflation and bursting tests: 100 of each were undertaken upon receipt of products, and the remaining 50 were tested following artificial ageing.
49. Comments on Draft Standard for Condoms, Washable Sheaths and Diaphragms, November 1958. WL/SA/FPA/A7/23.2.
50. Lamberts (Dalston) Ltd. Manufacturers Price List, [c.1951]. WL/SA/FPA/A7/23.1.
51. BSI and FPA Rubber Standard Informal Meeting Report, 15 September 1955. WL/SA/FPA/A7/23.2.
52. *Approved List*, December 1954. WL/SA/FPA/A7/5.
53. *Approved List Rubber Standard*, October 1954. WL/SA/FPA/A7/23.1.
54. *Approved List*, 1958. WL/SA/FPA/A7/5.
55. *Approved List*, 1958. WL/SA/FPA/A7/5.
56. *Ortho Pharmaceuticals Medical Methods of Conception Control Pamphlet*, April 1954. WL/SA/FPA/A7/23.1; Lamberts (Dalston) Ltd. Manufacturers Price List, [c.1951]. WL/SA/FPA/A7/23.1; Letter A. R. Reid to G. B. Carruthers, 24 September 1954. WL/SA/FPA/A7/23.1.
57. Answers to Occlusive Caps Size Questionnaire, 17 February 1956 (revised 14 March 1956). WL/SA/FPA/A7/23.2.
58. Letter H. A. Davidson to A. J. Clifford-Smith, 9 January 1956. WL/SA/FPA/A7/23.2.
59. Letter R. Harrison to I. James, 16 December 1954. WL/SA/FPA/A7/23.1.

60. Memorandum of FPA Visit to London Rubber, 15 September 1954. WL/SA/FPA/A7/23.1.
61. Report of FPA Medical Sub-Committee for Executive Committee, 22 February 1954. WL/SA/FPA/A7/23.1.
62. Borge, *Protective Practices*, 107.
63. Answers to Occlusive Caps Size Questionnaire, 17 February 1956 (revised 14 March 1956). WL/SA/FPA/A7/23.2.
64. Form Letter to Technical Committee Members, 15 January 1964. WL/SA/FPA/A7/23.2.
65. BSI Technical Committee on Surgical Rubber and Plastic Appliances Meeting Minutes, 31 October 1963. WL/SA/FPA/A7/23.2.
66. Form Letter to Technical Committee Members, 15 January 1964. WL/SA/FPA/A7/23.2.
67. Deborah A. Cohen, 'Private Lives in Public Spaces: Marie Stopes, the Mothers' Clinics and the Practice of Contraception', *History Workshop Journal* 35, 1 (1993): 95–116; Joyce M. Ray and F. G. Gosling, 'American Physicians and Birth Control, 1936–1947', *Journal of Social History* 18, 3 (Spring, 1985): 399–411; Hera Cook, 'Unseemly and Unwomanly Behaviour: Comparing Women's Control of Their Fertility in Australia and England from 1890 to 1970', *Journal of Population Research* 17, 2 (2000): 126.
68. Kate Fisher, *Birth Control, Sex and Marriage in Britain, 1918–1960* (Oxford: Oxford University Press, 2006), 55.
69. Consumers Association, *Contraceptives: A Which? Supplement* (England: Shenval Press Ltd., 1963), 2–3.
70. Letter M. Jackson to A. J. Clifford Smith, 1 February 1956. WL/SA/FPA/A7/22.
71. Note G. I. M. Swyer to FPA, 16 April 1959. WL/SA/FPA/A7/23.2.
72. Suzanne White Junod and Lara Marks, 'Women's Trials: The Approval of the First Oral Contraceptive Pill in the United States and Great Britain', *Journal of the History of Medicine* 57, 2 (2002): 128.
73. Letter FPA to D. A. W. Edwards, 10 September 1959. WL/SA/FPA/A7/22.
74. Final Draft Standard for Rubber Condoms, Washable Sheaths and Diaphragms, April 1960. WL/SA/FPA/A7/23.2; *British Standards Institution Specification for Rubber Condoms, B.S. 3704: 1964*, April 1964. WL/SA/FPA/A7/23.2. The only content difference was that the 1960 draft claimed 'articles selected [for testing] ... shall be at the discretion of the purchaser' and the actual standard clarified 'samples will be taken at random from each quantum of production'.
75. Letter B. Northage to D. Hills, 2 August 1960. WL/SA/FPA/A7/23.2; Memo M. Howard to E. Mears, 13 October 1961. WL/SA/FPA/A7/23.2.

76. Note from M. Howard, 23 October 1961. WL/SA/FPA/A7/23.2.
77. *British Standards Institution Specification for Rubber Condoms, B.S. 3704: 1964*, April 1964. WL/SA/FPA/A7/23.2; Letter from W. S. Hodges to R. C. Elestone, 31 July 1962. WL/SA/FPA/A7/23.2.
78. *Press Release BSI First Rubber Contraceptive Standard*, 8 April 1964. WL/SA/FPA/A7/23.2.
79. Draft British Standard for Diaphragms, December 1964. WL/SA/FPA/A7/23.2.
80. *Press Release BSI Second Rubber Contraceptive Standard*, June 1966. WL/SA/FPA/A7/23.2.
81. International Planned Parenthood Federation (IPPF) Meeting of Sub-Committee on Tests for Contraceptive Products, 21 November 1958. WL/SA/FPA/A10/15.
82. Birth Control Investigation Committee, *International Medical Group for the Investigation of Contraception, Fifth Issue* (London: Printed for the Birth Control Investigation Committee by George Standring, 1934), 14; H. Taylor, 'Report on the Hormonic Control of Fertility', in *The Practice of Contraception*, eds. Margaret Sanger and Hannah M. Stone (United States of America: Williams and Wilkins Co., 1931).
83. 'Contraception and the Profession', *The Journal of Contraception* 2, 2 (1937): 37.
84. 'Contraception and the Profession', 27.
85. Cedric Lane-Roberts, 'A Plea for the Woman in Gynaecology and Obstetrics', *Lancet* 1, 6027 (1939): 491–6; 'Kensington Medical Society', *Lancet* 2, 5963 (1937): 1375–7; 'Fellowship of Medicine and Post-Graduate Medical Association', *Lancet* 2, 5851 (1935): 910; 'The "Safe Period"', *Lancet* 1, 5977 (1938): 678–9; V. B. Green-Armytage, 'Sterility and Contraception', *British Medical Journal* 2, 4320 (1943): 524; Margaret Jackson, 'Contraception and Sterility', *British Medical Journal* 2, 4325 (1943): 691; Rafael Kuzrok, 'The Prospects for Hormonal Sterilization', *Journal of Contraception* 2, 2 (1937): 27–8; Allan Palmer, 'Excretion of Hormones in a Case of Habitual Abortion', *American Journal of Obstetrics and Gynecology* 36, 6 (1938): 1005–8; Hugo Ehrenfest, 'Review of *The Current Status of Female Sex Hormones*, by Roy G. Hoskins', *American Journal of Obstetrics and Gynecology* 44, 5 (1942): 915–6; Regine K. Stix, 'Research in Causes of Variations in Fertility: Medical Aspects', *American Sociological Review* 2, 5 (1937): 668–77; M. F. Ashley-Montagu, 'Adolescent Sterility', *Quarterly Review of Biology* 14, 1 (1939): 13–34; Clifford White, Comyns Berkley and Frank Cook, eds., *Diseases of Women*, Tenth Edition (London: Edward Arnold, 1938); Edward F. Griffiths, *Modern Marriage and Birth Control* (London: Victor Gollancz, 1935); Edward F. Griffiths, *Voluntary Parenthood* (London: Heinemann, 1937).

86. *International Medical Group for the Investigation of Contraception, Annual Report*, 1934. Wellcome Library, Archives of the Private Papers of Carlos Paton Blacker (WL/PPCPB), WL/SA/PPCPB/C/2.
87. F. H. A. Marshall, *The Physiology of Reproduction* (London: Longmans, Green and Co., 1910), 1.
88. Adele E. Clarke, *Disciplining Reproduction: Modernity, American Life Sciences, and the 'Problems of Sex'* (Los Angeles: University of California Press, 1998), 69–71.
89. V. C. Medvei, *The History of Clinical Endocrinology: A Comprehensive Account of Endocrinology from Earliest Times to the Present Day* (United Kingdom: The Parthenon Publishing Group, 1993), 218.
90. A. S. Parkes, 'Prospect and Retrospect in the Physiology of Reproduction', *British Medical Journal* 2, 5297 (1962): 72.
91. Response C. B. S. Hodson to BCIC and Eugenics Society, December 1934. Wellcome Library, Archives of the Eugenics Society (WL/SA/EUG), WL/SA/EUG/C/358.
92. Letter C. P. Blacker to S. Zuckerman, 6 January 1936. WL/SA/EUG/C/358; A. S. Parkes and Solly Zuckerman, 'Proposed Work on Fertility Period and De-nidation in Primates', October 1934. WL/SA/EUG/C/358.
93. Solly Zuckerman, 'Inhibition and Induction of Uterine Bleeding by Means of Oesterone', *Lancet* 228, 5888 (4 July 1936): 9–13.
94. Solly Zuckerman, 'The Menstrual Cycle', *Lancet* 253, 6564 (18 June 1949): 1031–5.
95. George W. Corner, 'Our Knowledge of the Menstrual Cycle, 1910–1950', *Lancet* 257, 6661 (28 April 1951): 919–23.
96. Corner, 'Knowledge of the Menstrual Cycle', 919–23.
97. Solly Zuckerman Lecture: 'The Physiology of Fertility in Man and Monkey', 21 January 1936. WL/SA/EUG/C/358.
98. This extrapolation is based on discussions of hormonal methods of contraception and sterilisation which first abounded in the 1930s and then reemerged in the 1950s. This position is derived from Wellcome Library Archives: WL/SA/EUG/C/358, WL/SA/EUG/D/168/173, WL/SA/FPA/A7/25A, WL/SA/PPCPB/C/2, and contemporary publications in the *Lancet, British Medical Journal*, and *Nature*.
99. 'Treatment of Involuntary Sterility', *Lancet* 243, 6302 (10 June 1944): 766–8.
100. Letter C. P. Blacker to S. Zuckerman, 1 March 1937. WL/SA/EUG/C/358.
101. Letter S. Zuckerman to C. P. Blacker, 3 June 1937. WL/SA/EUG/C/358.
102. Junod and Marks, 'Women's Trials', 125–8.
103. Consumers Association, *Contraceptives: A Which? Supplement*, 3.

104. Submission to British Pharmacopoeia Requesting an Official Contraceptive Preparation, 5 October 1953. WL/SA/FPA/A7/22.
105. 'The Family Planning Association', *British Medical Journal (Supplement 2205)* 1, 4501 (12 April 1947): 54.
106. 'The Family Planning Association', *British Medical Journal (Supplement 2205)*, 158.
107. 'The Family Planning Association', *British Medical Journal (Supplement 2205)*, 158.
108. Ministry of Health and Central Office of Information, *The National Health Service* (London: The Curwen Press, 1948), 1.
109. Ministry of Health and Central Office of Information, *The National Health Service*, 2.
110. Ministry of Health and Central Office of Information, *The National Health Service*, 6–8.
111. British Medical Association, *A National Health Service: Report of the Council of the BMA to the Representative Body* (St. Albans: The Gainsborough Press, 1944), 2.
112. British Medical Association, *A National Health Service*, 11.
113. Caroline Rusterholz, *Women's Medicine: Sex, Family Planning and British Female Doctors in Transnational Perspective, 1920–70* (Manchester: Manchester University Press, 2020), 49–53.
114. Harry Eckstein, *Pressure Group Politics: The Case of the British Medical Association* (London: Allen & Unwin, 1960), 93.
115. NBCA Executive Committee Meeting Minutes, 20 February 1948. WL/SA/FPA/A5/4.
116. Audrey Leathard, *The Fight for Family Planning: The Development of Family Planning Services in Britain 1921–74* (London: The Macmillan Press Ltd., 1980), 79.
117. Leathard, *Fight for Family Planning*, 224.
118. Borge, *Protective Practices*, 104–5.
119. FPA List of Approved Contraceptives Report, 1953. WL/SA/FPA/A7/16.
120. Daily Mail Newspaper Cuttings, 15 April 1958. WL/SA/FPA/A18/2; 'Contraception by Inhibition of Ovulation', *British Medical Journal* 2, 5091 (2 August 1958): 297; 'An Oral Contraceptive', *Lancet* 1, 7025 (1958): 845.
121. 'Contraception by Inhibition of Ovulation', 298.
122. Donald H. Merkin, *Pregnancy as a Disease: The Pill in Society* (London: Kennikat Press, 1976), 5.
123. Carl Djerassi, *This Man's Pill: Reflections on the 50th Birthday of the Pill* (New York: Oxford University Press, 2001), 54.
124. Junod and Marks, 'Women's Trials', 123.

125. Junod and Marks, 'Women's Trials', 123–4; Bernard Asbell, *The Pill: A Biography of the Drug That Changed the World* (New York: Random House, 1995), 118; Roy Porter, *The Greatest Benefit to Mankind: A Medical History of Humanity from Antiquity to the Present* (London: Fontana Press, 1997), 570.
126. Junod and Marks, 'Women's Trials', 125–8.
127. *FPA 30th Annual Report 1960–61*, 1961. WL/SA/FPA/A16/1/1.
128. John Peel, 'Contraception and the Medical Profession', *Population Studies* 18, 2 (1964): 134.
129. Jean Debelle, 'Hazards of the Pill', *The Australian Women's Weekly*, 18 February 1970.
130. Dugald Baird, 'A Fifth Freedom?', *British Medical Journal* 2, 5471 (13 November 1965): 1411.
131. 'The Contraceptive Pill', *British Medical Journal* 2, 5254 (1963): 754–5.
132. 'Oral Contraceptives and the NHS', *British Medical Journal* 2, 5353 (1963): 387; 'Discounts on Pharmaceutical Products', *British Medical Journal (Supplement)* 1, 5435 (1965): 77; Brian Kirkman, 'Contraception by Inhibition of Ovulation', *British Medical Journal* 1, 5340 (1958): 565–6.
133. Regine K. Stix, 'A Comparative Appraisal of Three Contraceptive Services', *Journal of the American Medical Association* 118, 4 (1942): 283–90; Michael S. Burnhill, 'The Intrauterine Contraceptive Device', *Journal of the American Medical Association* 195, 11 (1966): 970.
134. V. B. Green-Armytage, 'Contraceptives and Fertility', *British Medical Journal* 1, 4024 (1938): 419.
135. 'Gold Pin Pessaries', *British Medical Journal* 2, 4745 (1951): 249–50.
136. 'Fitting Dutch Caps', *British Medical Journal* 1, 5011 (1957): 181–2; Fred A. Simmons, 'The Cervical Cap', *Journal of the American Medical Association* 149, 12 (1952): 1156; 'Cap Contraceptives', *British Medical Journal* 1, 5236 (1961): 1405.
137. Eustace Chesser, 'The "Silver Ring"', *British Medical Journal* 2, 4480 (1946): 753; 'Gräfenberg Ring', *British Medical Journal* 1, 4773 (1952): 1418; 'Intrauterine Contraceptive Device', *British Medical Journal* 2, 4946 (1955): 1039; Rusterholz, *Women's Medicine*, 202, 214.
138. James Merchant, ed., *Medical Views on Birth Control, with an Introduction by Sir Thomas Horder* (London: Martin Hopkinson and Co. Ltd., 1926), 68.
139. Junod and Marks, 'Women's Trials', 128.
140. Gabriel V. Jaffé, 'Oral Contraception', *British Medical Journal* 2, 5244 (1961): 112; 'Contraindications to Oral Contraception', *British Medical Journal* 2, 5300 (1962): 315–6; 'Oral Contraceptives and the Liver', *British Medical Journal* 1, 5502 (1966): 1499.

141. 'The Pill Official Warning', *The Canberra Times*, 6 March 1970; 'The Pill', *The Australian Women's Weekly*, 30 July 1975.
142. Herbert Brewer, 'Are Oral Contraceptives Harmful?', *New Scientist* 344 (1963): 679–80; Asbell, *The Pill*, 110.
143. Note A. Parkes Visit to O. Bird, 15 March 1957. WL/SA/FPA/A5/128; FPA Establishment of the CIFC Meeting Minutes, 5 April 1957. WL/SA/FPA/A5/128; CIFC Project Position Reached, April 1937. WL/SA/FPA/A5/128; *Approved List*, April 1962. WL/SA/FPA/A7/11.
144. 'Oliver Bird Trust', Wellcome Library Digital Archive Record: WL/SA/FPA/A5/126–130, accessed 1 July 2022, http://archives.wellcome.ac.uk/DServe/dserve.exe?dsqIni=Dserve.ini&dsqApp=Archive&dsqDb=Catalog&dsqCmd=show.tcl&dsqSearch=(RefNo==%27SAFPA%2FA%2FA5%2F63%2F126%27).
145. Alan S. Parkes, 'The Oliver Bird Trust, 1957–69', *Journal of Biosocial Science* 2, 4 (October 1970): 359.
146. Junod and Marks, 'Women's Trials', 134.
147. Monica Mears, 'Obituary: E. Mears', *British Medical Journal* 305, 6850 (1992): 419.
148. Eleanor Mears, 'Clinical Trials of Oral Contraceptives', *British Medical Journal* 2, 5261 (1961): 1181; 'The Contraceptive Pill', *British Medical Journal* 2, 5254 (1961): 754–5; 'Oral Contraceptives', *British Medical Journal* 2, 5197 (1960): 551; G. B. Carruthers and G. I. M. Swyer, 'Oral Contraceptives Trials', *British Medical Journal* 1, 5289 (1962): 1418; 'Oral Contraceptives', *British Medical Journal* 1, 5179 (1960): 1145; P. N. Shah, 'Low-Dosage Oral Ethnydiol Diacetate Tablets for Long-Term Contraception in Indian Women', *British Medical Journal* 2, 5527 (1966): 1431–4; 'Overcrowded Asian Countries', *British Medical Journal* 2, 5244 (1961): 120; F. M. Purcell, 'Intrauterine Contraceptive Device and Population', *British Medical Journal* 2, 5472 (1965): 1242.
149. P. Eckstein et al., 'The Birmingham Oral Contraceptive Trial', *British Medical Journal* 2, 5261 (1961): 1173.
150. Denise Pullen, '"Conovid-E" as an Oral Contraceptive', *British Medical Journal* 2, 5311 (1962): 1017.
151. Mears, 'Clinical Trials of Oral Contraceptives', 1179–83; P. Eckstein et al., 'The Birmingham Oral Contraceptive Trial', 1177. In its first six months the original Birmingham clinical trial had a twenty-five per cent pregnancy rate. Following dosage recalculation, numbers dropped to less than one per cent.
152. Eleanor Mears and Ellen C. G. Grant, '"Anovlar" as an Oral Contraceptive', *British Medical Journal* 2, 5279 (1962): 78.
153. Junod and Marks, 'Women's Trials', 136.
154. CIFC Minutes, 1 October 1958. WL/SA/FPA/A5/154.

155. CIFC Minutes, 16 October 1958. WL/SA/FPA/A5/154.
156. Junod and Marks, 'Women's Trials', 137.
157. NKWWC Medical Committee Minutes, 9 February 1953. WL/SA/FPA/NK230.
158. *Sunday Pictorial*, 19 October 1958. Quoted in Adrian Bingham, *Family Newspapers? Sex, Private Life, and the British Popular Press, 1918–1978* (Oxford: Oxford University Press, 2009), 84.
159. *Sunday Pictorial*, 2 November 1958; *Sunday Pictorial*, 9 November 1958. Quoted in Bingham, *Family Newspapers?* 84.
160. 'Mass Trial of Birth Control Method', *The Times (London)*, 7 August 1959.
161. Pullen, '"Conovid-E" as an Oral Contraceptive', 1019.
162. P. Eckstein et al., 'The Birmingham Oral Contraceptive Trial', 1172–8.
163. Junod and Marks, 'Women's Trials', 147.
164. Junod and Marks, 'Women's Trials', 137–8.
165. Mears and Grant, '"Anovlar"', 76; P. Ritchie, 'Oral Contraceptives on the N.H.S.', *British Medical Journal* 2, 5358 (1963): 685; Pullen, '"Conovid-E" as an Oral Contraceptive', 1016–9.
166. Mears, 'Clinical Trials of Oral Contraceptives', 1182.
167. 'Sequential Oral Contraceptives', *British Medical Journal* 2, 5516 (1966): 747.
168. Leathard, *Fight for Family Planning*, 110–1.
169. Trent Stephens and Rock Brynner, *Dark Remedy: The Impact of Thalidomide and Its Revival as a Vital Medicine* (New York: Basic Books, 2001), 26.
170. Letter G. R. Venning to D. M. Shotton, 18 May 1960. WL/SA/FPA/A5/162/3; CIFC To Be Completed in the Case of an Unwanted Pregnancy Form, 1968–9. WL/SA/FPA/CB/5/1.
171. Letter N. M. Windette to O. C. Dobson, 30 September 1963. WL/SA/FPA/KN/14.
172. NKWWC Medical Committee Minutes, 24 March 1960. WL/SA/FPA/NK230.
173. NKWWC Medical Committee Minutes, 12 September 1961. WL/SA/FPA/NK230.
174. Dick Glover, 'Oral Contraception', *British Medical Journal* 1, 5223 (1961): 432.
175. Sandra Morgen, *In Our Own Hands: The Women's Health Movement in the United States, 1969–1990* (United States of America: British Cataloguing-in-production, 2002), 9–10; Lara Marks, *Sexual Chemistry: A History of the Contraceptive Pill* (London: Yale University Press, 2001), 117; Consumers Association, *Contraceptives: A Which? Supplement*, 86.
176. Consumers Association, *Contraceptives: A Which? Supplement*, 86.

177. *Medical Sub-Committee of Manchester Clinic Special Minutes*, 4 December 1962. WL/SA/FPA/X9/5.
178. Edward T. Tyler, 'Oral Contraception', *Journal of the American Medical Association* 175, 3 (1961): 225.
179. Bernard Notes, 'The Pill's Progress', *Journal of the American Medical Association* 207, 1 (1969): 154.
180. Marks, *Sexual Chemistry*, 133–4.
181. T. N. A. Jeffcoat and V. R. Tindall, 'Venous Thrombosis and Embolism in Obstetrics and Gynaecology', *The Australian and New Zealand Journal of Obstetrics and Gynaecology* 5, 3 (1965): 122; Heather Molyneaux, 'Controlling Contraception: Images of Women, Safety, Sexuality and the Pill in the Sixties', in *Gender, Health and Popular Culture*, ed. Cheryl Krasnick Warsh (Ontario: Wilfred Laurier University Press, 2011), 65–85; Fisher, *Birth Control, Sex and Marriage*, 239–40.
182. 'Oral Contraception', *Lancet* 279, 7240 (2 June 1962): 1167; Glover, '*Oral Contraception*', 432.
183. G. R. Venning, 'Contraindications to Oral Contraception', *British Medical Journal* 2, 5302 (1962): 478.
184. G. I. M. Swyer, 'Progestogens and Their Clinical Uses: Part 1', *British Medical Journal* 1, 5165 (1960): 49; G. Pincus, 'Field Trials with Norethynodrel as an Oral Contraceptive, n.d. WL/SA/FPA/A5/161/3. Quoted in Marks, *Sexual Chemistry*, 133; W. G. McBride, 'A Clinical Trial of a New Low Dose Oral Contraceptive Compound', *The Australian and New Zealand Journal of Obstetrics and Gynaecology* 7, 3 (1967): 222–3.
185. E. Mears, *'The Clinical Application of Oral Contraceptives' Paper Read to Symposium of Agents Affecting Fertility*, Middlesex Hospital Medical School, 24 March 1964. WL/SA/FPA/A5/15B. Quoted in Marks, *Sexual Chemistry*, 133.
186. Joint Meeting of the FPA Medical and Scientific Advisory Council and the CIFC Medical Trials Committee, 9 June 1970. WL/SA/CBS/1.
187. Junod and Marks, 'Women's Trials', 142; Jones, *The Business of Birth Control*; Borge, *Protective Practices*.
188. Eleanor Mears, 'A New Type of Oral Contraceptive', *British Medical Journal* 1, 5341 (1963): 1318–20; P. Eckstein et al., 'The Birmingham Oral Contraceptive Trial', 1175.
189. 'Oral Contraceptives', 2, 5179, 551; 'Contraindications to Oral Contraception', *British Medical Journal* 2, 5300 (1962): 315–6.
190. 'Oral Contraceptives and Thrombosis', *British Medical Journal* 2, 5301 (1962): 426.
191. NKWWC Medical Committee Minutes, 27 September 1962. WL/SA/FPA/NK230.
192. Letter M. Blair to R. Brain, 20 July 1962. WL/SA/FPA/NK230.

193. Letter E. Mears to M. Blair, 30 July 1962. WL/SA/FPA/NK230.
194. Michel Foucault, *The History of Sexuality, Vol. 1: An Introduction* (New York: Pantheon Books, 1978), 99.
195. J. Shafar and G. Behr, 'Oral Contraceptive and Thrombophlebitis', *British Medical Journal* 2, 5318 (1962): 1543; F. Nour-Eldin, 'Oral Contraceptives and Blood Coagulability', *British Medical Journal* 1, 5327 (1963): 400; L. O. Pilgeram, 'Blood Coagulability and Oral Contraception', *British Medical Journal* 1, 5387 (1964): 883–4; Reginald T. Payne, 'Oral Contraceptives, Thrombosis, and Cyclical Factors Affecting Veins', *British Medical Journal* 1, 5478 (1966): 51; L. Poller, Anne Tabiowo, and Jean M. Thomson, 'Effects of Low-Dose Oral Contraceptives on Blood Coagulation', *British Medical Journal* 3, 5612 (1968): 218–9; Ellen C. G. Grant, 'Venous Effects of Oral Contraceptives', *British Medical Journal* 4, 5675 (1969): 73–7; J. Scorey, 'Oral Contraceptives and Thrombosis', *British Medical Journal* 2, 5456 (1965): 301; Jeffcoat and Tindall, 'Venous Thrombosis and Embolism', 120–1.
196. J. G. Scott, 'Oestrogens and Thrombosis', *British Medical Journal* 2, 5356 (1963): 558.
197. Pilgeram, 'Blood Coagulability', 883–4; Scorey, 'Oral Contraceptives and Thrombosis', 301; McBride, 'New Low Dose Oral Contraceptive', 222; W. A. W. Walters, 'Cardiovascular Disease Contributing to Maternal Mortality in Victoria', *The Australian and New Zealand Journal of Obstetrics and Gynaecology* 9, 1 (1969): 1–6.
198. Edward T. Tyler, 'Eight Years Experience with Oral Contraception and an Analysis of Use of Low-Dosage Norehisterone', *British Medical Journal* 2, 5413 (1964): 845.
199. Jeffcoat and Tindall, 'Venous Thrombosis and Embolism', 122; 'Oral Contraceptives and Thromosis', *British Medical Journal* 2, 5400 (1964): 38.
200. Letter T. Fox, to J. Roos, 4 June 1965. WL/SA/FPA/A7/29.
201. 'Legal Contraindications to Oral Contraceptives', *British Medical Journal* 2, 5553 (1967): 709; Hugh O. Jones, Julian C. F. Townsend and J. T. Roberts, 'Varicose Veins, Oral Contraceptives, and Thromboembolism', *British Medical Journal* 2, 5552 (1967): 637–8; 'Oral Contraceptives and Thrombo-Embolic Disease', *British Medical Journal* 2, 5548 (1967): 327–8; G. A. Holbrook, 'Oral Contraceptives', *British Medical Journal* 2, 5545 (1967): 174.
202. North Kensington Marriage Welfare Centre Medical Committee Minutes, 24 March 1960. WL/SA/FPA/NK230; Manchester Medical Sub-Committee Minutes, 21 September 1961. WL/SA/FPA/X9/5.
203. Donald R. Lavis, *Oral Contraception in Melbourne: An Investigation of the Growth in Use of Oral Contraceptives and Their Effect Upon Fertility*

in Australia, 1961–1971, Australian Family Formation Project, Monograph No. 3 (Canberra: Department of Demography, Australian National University, 1975), 26; 'The "Pill" and Thrombosis', *British Medical Journal* 2, 5417 (1964): 1089.
204. North Kensington Marriage Welfare Centre Medical Committee Minutes, 25 March 1965. WL/SA/FPA/NK230; Manchester Medical Sub-Committee Minutes, 25 October 1955. WL/SA/FPA/X9/5; Marie Stopes Clinic Medical Sub-Committee Minutes, 30 October 1964. WL/SA/EUG/K44.
205. Matthew Connelly, *Fatal Misconception: The Struggle to Control World Population* (Massachusetts: Belknap Press of Harvard University Press, 2008), 270.
206. Note of an Informal FPA Meeting following a Meeting of the Oliver Bird Trustees, 29 July 1957. WL/SA/FPA/A5/162/5.
207. Junod and Marks, 'Women's Trials', 151–2.
208. Junod and Marks, 'Women's Trials', 136.
209. 'Oral Contraceptives', *Lancet* 2, 7144 (1960): 256.
210. NKWWC Medical Committee Minutes, 7 November 1963. WL/SA/FPA/NK230.
211. NKWWC Medical Committee Minutes, 17 January 1963. WL/SA/FPA/NK230.
212. NKWWC Medical Committee Minutes, 12 September 1961. WL/SA/FPA/NK230; Manchester Medical Sub-Committee Minutes, 23 March 1964. WL/SA/FPA/X9/5.
213. Report on the Sphere of Work for FPA Doctors, 1 December 1965. WL/SA/FPA/A19/14.
214. Junod and Marks, 'Women's Trials', 152.
215. NKWWC Medical Committee Minutes, 7 November 1963. WL/SA/FPA/NK230.
216. E. M. Tansey, *Wellcome Witness to Twentieth Century Medicine, Volume 1, Technology Transfer in Britain: The Case of Monoclonal Antibodies; Self and Non-Self: A History of Autoimmunity; Endogenous Opiates; The Committee on Safety of Drugs, Witness Seminar Transcripts* (London: Wellcome Trust Centre for the History of Medicine, 1997), 107.
217. 'Oral Contraceptives and Cancer', *Lancet* 2, 7783 (1972): 911.
218. Elizabeth B. Connell, 'The Pill Revisited', *Family Planning Perspectives* 7, 2 (March/April 1975): 63–4.
219. Antti Eisalo, Pentti A. Järvinen and Tapani Luukkainen, 'Hepatic Impairment During the Intake of Contraceptive Pills: Clinical Trial with Postmenopausal Women', *British Medical Journal* 2, 5406 (1964): 426–7.

220. I. P. Palva and O. O. Mustala, 'Oral Contraceptives and Liver Damage', *British Medical Journal* 2, 5410 (1964): 688–9; 'Oral Contraceptives and Liver Damage', *British Medical Journal* 2, 5411 (1964): 755.
221. G. Linthorst, 'Liver Function After Long-Term Progestational Treatment with and Without Oestrogen', *British Medical Journal* 2, 5414 (1964): 920–1; John L. Bakke, Hepatic Impairment During Intake of Contraceptive Pills: Observations in Post-Menopausal Women', *British Medical Journal* 1, 5435 (1965): 631–2; 'Oral Contraceptives and Liver Injury', *British Medical Journal* 1, 5447 (1965): 1391.
222. Junod and Marks, 'Women's Trials', 137; Screening Test for Oral Contraceptives, 3 August 1960. WL/SA/FPA/A5/162/3; CIFC Doctors Conducting Clinical Trials Minutes, 18 October 1965. WL/SA/FPA/A5/162/5; Letter N. M. Windett to H. Nelson-Barrett, 16 January 1963. WL/SA/FPA/A5/162/2; List of Requirements for Volunteers for Barnet Oral Contraceptive Trial, [c.1964]. WL/SA/FPA/A4/A1/3.
223. 'Oral Contraceptives and the Liver', *British Medical Journal* 1, 5502 (18 June 1966): 1499.
224. Djerassi, *This Man's Pill*, 72; A. Arnaud Reid, 'Oestrogens and Thromboembolism', *British Medical Journal* 4, 5625 (1968): 255.
225. 'Safety of Contraceptive Pills', *British Medical Journal* 2, 5419 (1964): 1276; 'Government to Inquire into Pill Side Effects', *The Canberra Times*, 10 November 1964.
226. 'Reports Expected on the Pill', *The Canberra Times*, 23 April 1968; Leathard, *Fight for Family Planning*, 111.
227. FPA Medical Sub-Committee Minutes, 17 January 1964. WL/SA/EUG/K44.
228. FPA Medical Sub-Committee Minutes, 30 October 1964. WL/SA/EUG/K44.
229. FPA Medical Sub-Committee Minutes, 30 November 1967. WL/SA/EUG/K44.
230. FPA Medical Sub-Committee Minutes, 3 May 1965. WL/SA/FPA/X9/5.
231. Draft Letter CIFC to G. G. Robertson, December 1963. WL/SA/FPA/A5/162/1.
232. 'Risk of Thromboembolic Disease in Women Taking Oral Contraceptives', *British Medical Journal* 2, 5548 (1967): 355–9; 'Oral Contraception and Thrombo-Embolic Disease', *Journal of the College of General Practitioners* 13, 3 (1964): 267–79.
233. 'Oral Contraception and Thrombo-Embolic Disease', 267.
234. 'Risk of Thromboembolic Disease', 357–9.
235. Tansey, *The Committee on Safety of Drugs*, 126; Junod and Marks, 'Women's Trials', 157.

236. 'Impasse on Birth Control Pill', *The Times (London)*, 29 August 1969; 'Birth Pill Killed Wife', *The Times (London)*, 16 October 1969; 'Death Risk from Birth Control Pill "Inevitable"', *Daily Telegraph*, 17 November 1969; 'Panic, Prudence and the Pill', *Observer*, 14 December 1969; 'Pill Makers Fight Back', *Evening Standard*, 12 December 1969; 'It's the Best Pill We've Got', *The Times (London)*, 13 December 1969.
237. Leathard, *Fight for Family Planning*, 112.
238. '*American Experience: The Pill*', Public Broadcasting Service Television Program Transcript, accessed 1 July 2022, http://www.pbs.org/wgbh/americanexperience/films/pill/#transcript.
239. Suzanne White Junod, 'Women Over 35 Who Smoke: A Case Study in Risk Management and Risk Communications, 1960–89', in *Medicating Modern America: Prescription Drugs in History*, eds. Andrea Tone and Elizabeth Siegel Watkins (New York: New York University Press, 2007), 104–5.
240. Leathard, *Fight for Family Planning*, 108.
241. FPA Medical Sub-Committee Minutes, 8 May 1973. WL/SA/EUG/K44.
242. A. S. Parkes, 'Control of Human Fertility: Introduction', *British Medical Bulletin* 26, 1 (1970): 1–2.

CHAPTER 6

Conclusion: The Fittest Survived?

Abstract As sex and contraception became ever more open to public debate, contraception itself became less overt. Its only real physical impact was internal—and thus imperceptible outside its host. With the official medical and scientific oversight of contraceptives accomplished, it was the individual biological regulation of synthetic hormones that now warranted focus. The conclusion will discuss the extent to which hormonal contraceptive technology and science intersected in the lab, the clinic, the bedroom and the body. And will ultimately answer the question: did the 'fittest' contraceptives survive?

The oral contraceptive revolution advanced such a comprehensive transition to the methods and means of fertility control that the older contraceptive technologies that had been refined and flourished for decades, and even centuries previous, initially fell from favour. The ease and convenience of a daily pill for women that theoretically didn't cause the body to function any differently from how it would on any other given day and provided a completely imperceptible (to the outside world), almost certain protection from the risk of pregnancy, seemed almost too perfect a contraceptive. Unlike the messy, sticky, fiddly, confusing, insertable, fallible and explicit technologies that were the subject of the majority of this book, the pill was profoundly different in every respect—it achieved all of the stated goals of the original union of contraceptive activists and

N. Szuhan, *The Family Planning Association and Contraceptive Science and Technology in Mid-Twentieth-Century Britain*, Medicine and Biomedical Sciences in Modern History, https://doi.org/10.1007/978-3-030-81300-0_6

scientists to create a contraceptive that was 'nontoxic, nonstaining, cheap, tolerant of storage under diverse environmental conditions [and] easy to use'.[1] Further, it was further asexual, small and simple, completely female-managed and nearly infallible. In short, by any early 1960s Darwinian comparison, the pill had evolved to become the 'fittest' contraceptive on the market—it was developed from the most important and useful aspects of the contraceptive science and medicine that preceded it, to create a truly efficient and easy contraceptive for the masses.[2] It should be noted that nature (in a conventional sense) had no hand in determining what was and was not vital for and to this ideal contraceptive. Nor had the women who would ultimately apply the drug; only male scientists and physicians did. Hormonal contraceptive methods were able to be refined due to a scientific recalibration of women. Until very recently, women were seen as complete patients with diverse and divergent sexual, reproductive and contraceptive needs that were broad and autonomous. The pill reduced this holistic view of women to simply a menstrual cycle. As Nelly Oudshoorn has argued, for the doctors and scientists involved in laboratory, field and clinic trials this conceptual revision was logical and straightforward, but 'disciplining women to submit themselves … to these examinations' and ideas 'was not an easy task'.[3] Women were essentially being asked to view their sexual and reproductive selves as being subordinate to their hormonal cycles. For many young women who had reached sexual maturity in the era of the pill, this was acceptable and perhaps even ideal—guaranteed protection was a matter of hormones, drugs and maths. But for many older women, even those involved in early global oral contraceptive trials, the product and its method of action were less ideal than a surgical sterilisation, and indeed many returned to some of the older methods pioneered and proved by the National Birth Control Association (NBCA)/Family Planning Association (FPA) from the 1930s.

The pill actually represented a contemporary 1960s version of the sociological contraceptive problems the association had faced since its inception: the challenge to convince women to submit themselves to the scientific contraceptive methods and advice tested, advocated and prescribed by its medical and scientific staff. But the socio-cultural environment was greatly changed, in no small part due to the association and its family planning and education-focused networks, as well as universal aspirations to lifestyle standards and mores that were contingent on small families, and marriages that prioritised sexual compatibility and romantic love. By the time the pill was rolled out in Britain, there was a female-focused and led push for a more assertive and equality-based femininity that accepted medical and scientific interventions in the realms of sexuality and fertility as far as they promoted self-determination in each of these arenas. It was this sense of educated contraceptive autonomy, that the association had helped to foster and imbue in younger generations of British women, that facilitated trust in medicine and science to provide couples ever 'more desirable [technological] solution[s] to the problem of fertility control' that further liberated sex from reproduction as the century progressed.[4] The legacy of the 1974 FPA and National Health Service union through the NHS Reorganisation Act that took effect that year, was that couples could then easily access fully subsidised officially- and scientifically-guaranteed medical contraceptives.[5] This legislation ensured that all contraceptive methods that required medical fitting and prescription, were able to be widely publicised and their methods of application and operation taught by trained practitioners in doctors rooms, hospitals and contraceptive clinics, as well as being easily and financially accessible for all. With the introduction of the Act, the association succeeded in ensuring that women could access verified technologies and therapies that suited their personal, sexual, educational and reproductive capacities and needs.

Nonetheless, the advent of the pill and the contraceptive accessibility and permissibility that followed, prompted social and cultural changes that significantly impacted British society in much broader ways than even the most strident FPA advocate ever anticipated, especially regarding marital relations and demographics. Anthropologist Susan Scrimshaw claimed that the 'pill ha[d] been a catalyst for…[the] most rapid and profound changes in sexual attitudes and behaviour, in the altered dynamic of the male–female relationship, [and] in the role of women in society'. The specific causality of the drug for marital and birth rate decline, as well as its role in sexual and women's liberation more broadly, has been questioned by some social and actual scientists over the last half century; but the effective transition of contraceptive control from old wives tales and recipes, to men through strategic and skillful application of coitus interruptus and condoms, to scientists and physicians, and finally to female users, undoubtedly altered women's relationship to sex, and thus men, family and society.

Dilys Cossey, the FPA Chairwoman from 1987 to 1993, explained the radical social, economic and cultural impact of oral contraceptives.

> The concept ... that women could just put something in their mouths and were safe... profoundly challenged men because if you have an independent, sexual being she can behave in any way she wants without being controlled.

She continued that as with other 'big new developments' like industrialisation, antibiotics, nuclear weapons, air and space travel and now automation, 'new problems start asserting themselves'.[6] The pill freed women from some problems associated with sex and fertility, but it facilitated others. Having the freedom to avoid the inherent biological ramifications of coitus was not the key to female happiness, rather women were 'given the power to plan their lives', they were 'forced to grow up' and individually choose when and if to have sex and a family—but this choice did not guarantee happiness or that their choices would be correct.[7]

The Family Planning Association's successful efforts to design, implement and normalise standardised contraceptive education, methods, and products were ultimately progressed using the scientific and medical advances of the era and tapped into the existing and emerging spoken and unacknowledged sexual, marital and family planning needs of several generations of Britons. But they did not profess to be about social or cultural transition outside this realm. The association promoted a firm policy of self-determined population control to meet the autonomous needs of each individual patient and user, as I have shown, there was certainly population control and eugenic ideologies that simmered at the periphery of the NBCA/FPA function and propaganda, but this remained intentionally tangential to its overt medico-scientific and regulatory agendas. The association's management evolved and adapted its practices over the forty years under discussion to transition from a proactive welfare organisation to a reactive medico-scientific and therapeutic service provider that aligned its methods and practices with up-to-date medical developments and responded to emerging social and sexual needs—such as the provision of contraceptives to the unmarried, although it was relatively slow to implement that particular service.

By the era of Cossey's FPA leadership, the relationship of contraception to marriage and sex had transitioned significantly. By 1960 the public link between contraception and marriage was no longer exclusive, as Jessica Borge has shown there was a tension that had emerged between older barrier contraceptives like the condom and the modern, medicalised pill.

The latter was associated with science, progress and a positive and desirable overhaul of sexual attitudes and mores; whilst condoms had long been used as a marital contraceptive aid, but remained widely and pejoratively associated with casual, unmarried or extramarital sexual relations.[8] Oral contraceptives would be the panacea for the put-upon wife and mother whose family was already large enough or required guaranteed spacing but didn't wish to compromise her sexual and marital relationship through abstinence or the risk of an unwanted pregnancy. However, despite the comparatively *longue durée* of the FPA's firm promotion of marital contraceptive norms for the mechanical and chemical methods that it had been dedicated to for decades, its restriction on the pill to married women was very short-lived. By 1971 the association was publicising that it would 'prescribe the pill for unmarried girls providing there [wa]s no medical reason why [it] shouldn't'.[9] This was a huge transition in the FPA's stated attitude to its standards of contraceptive provision, as in its 1963 report *Family Planning in the Sixties*, it remained adamant that 'sexual activity amongst the unmarried' would facilitate some 'specific social problems' including promiscuity, prostitution, divorce and non-married immigrants being brought to the country through common-law marriages. The FPA had long argued that it could not be seen to be 'encouraging immorality' and thus reiterated its long-held mantra that its primary duty was to advise on and provide means of 'fertility regulation within the context of family life – that is to advise the married and those about to marry'.[10] Yet, the next year former FPA affiliate Lady Helen Brook opened the first Brook Advisory Centre in London to specifically provide contraceptive advice and prescriptions to unmarried women and girls under the age of twenty-five. By March 1969, the FPA was in the same game and provided space in its new Tottenham Court Road premises for another site to meet this obvious need.[11]

This new site marked the association's first admission that a significant sexual and social transition had taken place in Britain—the millennia-old male–female sexual balance had been turned on its head and there was little doubt that the advent of an effective asexual contraceptive was a prime cause of changes to the variety, age and frequency of sexual activity in which Britons engaged. This transition, first dubbed the 'permissive moment' by historian Jeffrey Weeks in the 1980s and later more universally defined as the 'sexual revolution', was undoubtedly a direct product of the entirely correct common understanding that the drug was practically infallible and the resultant alteration of 'people's expectations about contraception and what it could achieve'.[12]

The association's multitudinous efforts to ensure that the mechanical, chemical and later hormonal contraceptives that it made available in its clinics could be guaranteed as safe, efficacious and standardised, promoted a faith in the practice and methods of contraception that by the time of the pill promised to guarantee temporary sterility through pharmaceutical means. This advent had a staggered effect, initially it made older and less certain technologies seem even more outdated, ineffective, messy and undesirable, but also it meant that when the pill's safety was called into question, there were a multitude of more 'natural', safe and effective contraceptives immediately available that were female-controlled and thus perpetuated the cultural shift towards women-focused social movements into the 1970s. In the post-pill era, three cohorts of women attended family planning association clinics and general practitioners' rooms specifically to obtain contraceptives: 'young, single nulliparous women' delaying first pregnancy, 'older, parous women' with complete families, and 'certain professional women'. This female-focused and directed cultural shift had long been the goal of the FPA and its pro-contraception partners and competitors. In the 1960s the desired cohort of fertile women had congealed into a proactive, educated and politicised bloc that was well informed about sex and contraception and prioritised and trusted FPA and regulatory efficacy guarantees over potential and unlikely safety threats.[13] Indeed in this new contraceptive climate many of these women reported to have taken 'the unusually assertive move' of changing their doctors if one physician refused to prescribe them desired contraceptives—or denied them access to the NHS subsidy.[14] In a 1985 Royal College of General Practitioners survey of contraceptive use amongst sexually active British women between 16 and 29 years of age, only five per cent of respondents reported that they had never tried the pill and those in older and lower class cohorts were only slightly less likely to have tried the drug.[15] This represented an overwhelming embrace of contraceptive technology within half a century, and also an unmistakable indication that the FPA, contraceptive manufacturers and retailers, women doctors, scientists and users were driving a distinct contraceptive cultural change which was based on the professionalisation and regulation of methods and products that had taken place since the 1920s.

The pill really was a revolutionary contraceptive technology that had huge social, political, religious, medico-scientific and autonomy ramifications. That is the primary reason why it holds such an important and well-studied pride of place in the contraceptive history canon. But as I have shown, the medical, scientific, propagandistic and educational groundwork for these changes was laid over the previous half century by, mostly female, NBCA/FPA contraceptive advocates, activists, and scientists and physicians, who in affiliation with the Birth Control Investigation Committee, the Society for the Provision of Birth Control Clinics, Marie Stopes' and her affiliates, and the Eugenics Society and other family-focused education and therapy groups promoted and guaranteed the practice as a medical speciality. The association's medicalised facilities, practices and standards were key to effecting the conceptual and pragmatic changes that made the drug such an appealing technology for users and providers. The fallibility of hands-on, individually fitted and applied devices that had been a major concern of the FPA from a sociological perspective, was effectively rendered moot by the oral contraceptive that required daily ingestion at a consistent time, rather than an intimate and correct knowledge of one's own physiology and a keen enough interest in pregnancy prevention to apply diaphragms/caps or condoms and sheaths and their complementary spermicides prior to each intercourse.

Hera Cook identified the period between 1965 and 1970 as the years of the greatest social change.[16] This half decade saw the number of pill users in Britain rise from 480,000 in 1964 to one million in 1969[17]; by 1975 the number had risen to over two and a quarter million.[18] In addition, a London Rubber-funded study conducted by the University of Hull's Sociological Research Foundation found that male-controlled methods—namely the condom—were then being used at twice the rate of the pill.[19] So, there was a widespread openness to contraceptive use and a duopoly of scientifically and sociologically proven gendered methods; an asexual but medically managed and regulated method was preferred by female consumers/patients and the mechanical male-controlled and applied condom remained a staple and its market continued to grow annually. This was concurrent with a raft of 'government ushered… liberalising measures – partially legalising abortion and homosexuality in 1967 and introducing no-fault divorce in 1969'.[20] These liberalisations have been linked to the pill in the assessments of social scientists, politicians, clergymen, feminists, and users—and the link was not exclusive to the UK and the United States, in fact it drove all of the societies where the product was introduced and consumed en masse towards sexual openness and 'autonomy'.

This uptake of the pill coincided with the initiation of the Women's Liberation Movements which first formed in Western European nations in the wake of the late 1960s student and workers movements, and really took off in the 1970s. Its adherents were those young women that had embraced oral contraceptives and sexual emancipation in the naive belief that the pharmaceutical and medical industries were benevolent technocrats bearing earnest gifts. But as this work has discussed, the FPA at its core operated from a top-down perspective. It worked to formalise a centralised management and cascading bureaucracy that instituted and oversaw a standardised set of practices that could be trialed and checked through its clinics and practices. It was from this position of power and control that the association approached and ministered the needs of its patients. It was absolutely intended as a benevolent service, but also sought to be seen and approached as a technocratic authority. This is evident in its self-standardising and regulating practices; it was consistently dedicated to encouraging both compliance and reliance, in that it was a primary provider of female liberation from the threat of pregnancy, as well as other sexual and maternal risks, but this could only be achieved by faithfully following its contraceptive prescriptions.

But the proposed remedy to the deep history of female sexual and social repression that was now being identified and challenged within the movement—an infallible contraceptive technology like the pill—proved a poisoned chalice.[21] Oral contraceptives were revolutionary and did cause significant changes to male–female relations in terms of partnerships and sex, but it was not a panacea for all the woes of women living in a patriarchal society. It merely reorganised the traditional social and sexual labours of women to facilitate their mass entry into the workplace in addition to their traditional home duties, and through the guarantee of sex without the consequence of pregnancy, enabled some men to enjoy continued sexual dominance by removing a significant but by then outmoded reason for female reserve. In fact, men increasingly assumed it was the woman's sole responsibility to obtain and ingest oral contraceptives, and thus diminished the masculine sense of responsibility that Kate Fisher identified as a key ideal and practice of companionate marriages in the immediately preceding decades.[22] Although the pill did provide an opportunity for some women to experience unfettered sexual pleasure, women were increasingly perceived to be solely responsible for its facilitation and achievement—even if their sexual desires were not for 'sex as a hedonistic act, for pleasure or intimacy' as was often charged.[23] Cook succinctly described the situation of many women at this time; 'In gaining the possibility to say yes' to pre- or post-marital sex through effective technological means 'women lost the right to say no'. This situation was merely a reversal of conventional male–female sexual power dynamics wherein women never really did have the authority to approve or decline sex based on their own desires; but initially, the pill created an impression that women could express and pursue their autonomous sexual aspirations.[24] This was never the goal of the association in its research or teaching, it did aim to provide some sexual autonomy for women, but coded its approach in the language of responsible family planning and building. Sex was, in its estimation and approach, a matter of marital fortification to ensure the strongest parental bond for future child-raising, it ought not to be gendered but also was not really an exclusive matter of sexual pleasure for its own sake.

Both contrary to and in service of this family focus, hormonal contraceptives also transformed the nature of marital and family units. This method finally achieved the FPA's goal of creating an infallible contraceptive and this ensured young couples could be absolutely certain of the effectiveness of applied contraceptives. The initial social effect was the validation first of married couples, and later unmarried couples, as legitimate and viable social units—children, which the association consistently reminded the public were always at the forefront of its focus and ideals, hence became somewhat peripheral to the concept of partnership and family. The ultimate result was that the 'responsible parenthood' doctrine that the FPA had advocated since its inception became the norm, and prospective parents increasingly delayed marriage and childbearing—resulting in declining birth and marriage rates, smaller families, and increasing divorces, numbers of children born out of wedlock and single parent families.[25]

Despite these social, sexual and familial compromises, many women still chose to take the pill as a way to control their reproductive bodies. The perception of control and thus choice and autonomy that modern hormonal and mechanical contraceptive technologies guaranteed, provided women a sense of control, autonomy and security and a limitation on each of these guarantees; women were and still are herded into its use, and as most pharmaceuticals must, this one-size-fits-all contraceptive has effectively dominated the market by 'reducing everyone to his or her common denominators'. Although women gained the sense and experience of finally being in ultimate control of their bodies, sex and families, actually they were experiencing 'dependent determinism'. Therein through a belief that they chose to take the pill as their primary contraceptive and their engagement with the medico-scientific trial and error processes invariably associated with finding the most suitable variety of pill for each user, women as a whole experienced a 'reduce[d] variation in [contraceptive] experience' through the drug's provision of 'a highly controlled and standardized cycle'.[26]

Users of contraceptive technologies throughout the twentieth century have always had a tenuous relationship with the concept of choice—primarily, in terms of method and need. And as this book has shown, the Family Planning Association was very much entrenched in controlling and directing those methods and needs. The pill did not resolve all of the long-term issues of 'choice', 'control' and 'freedom' associated with contraception and reproduction, but it did transform them. These three descriptors were increasingly applied to contraception in private and public discourse, whether they were applied to users, approving or disapproving partners, doctors, religious and political institutions, and family planning advocates. And through increased public discourse and exposure to the pill, the practice of contraception was finally liberated from the laboratory, clinic and bedroom, and brought into community, educational and medical focus.

But did this technology and its standardisation and regulation lead to positive socio-cultural change? Was this the brave new world that the original birth controllers, or even the 1950s family planners, envisioned when they formed voluntary social and scientific collectives to bring contraceptive technologies to the masses? The answer is complicated.

In the 'permissive society' that formed in the UK during the swinging sixties, FPA activists, scientists and physicians lagged behind more modern and progressive groups focused on related issues like abortion and divorce[27]; however, between 1930 and 1969 the association and its members had been prime in facilitating social and sexual liberation through their efforts to ensure only the 'survival of the fittest in contraceptives'.[28] The breadth and speed of social, cultural, medico-scientific, sexual, religious, political and legislative changes that followed the arrival of oral contraceptives were unpredictable, but as this book has shown, many of the changes that occurred were welcomed and anticipated by the association. Female sexual liberation, education and satisfaction were all overt goals; as was the ability for couples to enjoy sex as a non-procreative act; and the liberalisation of contraceptive access and oversight through official government medical channels. And ensuring no prohibitive cost for the service and products.

But in the era of oral contraceptives, the FPA adhered to its tradition of staying the course and remaining internally focused in the face of radical change, just as it had when US-based scientists first embraced the contraceptive potential of hormones. During the 1960s and early 1970s the association continued to focus on its safety and efficacy work and the extension and expansion of its clinic services.[29] It also steadfastly continued to argue in favour of a dual mechanical and chemical method such as a condom or cap and spermicide, as the best-proven methods. It enjoyed some success in this agenda when many British women returned to or embraced the diaphragm when the 'pill kills' campaign led them to abandon oral contraceptives. But by the 1970s there was also a greater demand for long-term, low-maintenance, asexual methods, and so it embraced and promoted intra-uterine devices which then enjoyed a resurgence in popularity, as did vasectomy and sterilisation which overwhelmed NHS facilities with over 150,000 operations being performed between October 1977 and 1978 and ever-increasing waiting lists for the services.[30]

Mechanical and chemical technologies required motivation, consistency and personal responsibility to guarantee efficacy, whereas the pill and other modern hormonal and long-term reversible contraceptive technologies that followed required no similar dedication. What these new methods did achieve was the creation of complacency in the face of new, proven technologies. A similar phenomenon is currently occurring with the internet; it has provided a means for users to immediately gratify a desire for knowledge via a well-constructed google search. As a result, libraries and archives are closing or being selectively digitised, attention spans are much reduced and the ability of the masses to master the complexities of issues and techniques is dwindling. Oral contraceptives made physicians, scientists and government regulators the responsible parties in the equation, and by removing the requirement for personal responsibility and technique in contraceptive application, engendered a belief that 'women using oral contraceptive pills were being used as guinea pigs' in a mass experiment.[31] This was of course true, but the claim depletes the agency of women who clearly researched and understood the risks of the pill, and through informed consent chose to employ the drug.

Ultimately for the pharmaceutical, medical and scientific communities, and contraceptive users, oral contraceptives in spite of their well-known risks, were deemed the most superior and convenient technology available in late twentieth-century Britain. As one early user succinctly clarified 'the pill isn't just the best available contraceptive method; it is also in a personal sense an emancipator of women',[32] whilst this may be true for many, even most users, the perfect contraceptive technology remains elusive.

Notes

1. Merriley Borell, 'Biologists and the Promotion of Birth Control Research, 1918–1938', *Journal of the History of Biology* 20, 1 (1987): 72.
2. William Bynum, *A Little History of Science* (London: Yale University Press, 2013), 146–53.
3. Nelly Oudshoorn, *Beyond the Natural Body: An Archaeology of Sex Hormones* (London: Routledge, 1994), 129.
4. Annette B. Ramírez de Arrellano and Conrad Seipp, *Colonialism, Catholicism and Contraception: A History of Birth Control in Puerto Rico* (Chapel Hill: University of North Carolina Press, 1983), 134.
5. Family Planning Association, *Factsheet: A History of Family Planning Services* (London: Family Planning Association, 2011), accessed 1 July 2022, https://www.fpa.org.uk/factsheets/history-family-planning-services#timeline.
6. Linda Grant, *Sexing the Millennium: A Political History of the Sexual Revolution* (London: HarperCollins Publishers, 1993), 59.
7. Grant, *Sexing the Millennium*, 59.
8. Jessica Borge, *Protective Practices: The London Rubber Company and the Condom Business* (Montreal & Kingston: McGill-Queen's University Press, 2020), 218–9.
9. FPA Pamphlet: The Pill, 1971. Wellcome Library, Archives of the Abortion Law Reform Association (WL/SA/ALR), WL/SA/ALR/U48.
10. Family Planning Association, *Family Planning in the Sixties: Report of the Family Planning Association Working Party, September 1963* (London: Family Planning Association, 1963), ch. 7, 17.
11. 'London Brook Centres', *FPA News*, March 1969. WL/SA/ALR/U48.
12. Lara V. Marks, *Sexual Chemistry: A History of the Contraceptive Pill* (New Haven: Yale University Press, 2001), 3.
13. Sam Rowlands, 'Family Planning: General Practice and Clinic Services', *Journal of the Royal College of General Practitioners* 35, 273 (1985): 199.

14. Hera Cook, 'The English Sexual Revolution: Technology and Social Change', *History Workshop Journal* 59, 1 (2005): 117.
15. John Guillebaud, *The Pill* (Oxford: Oxford University Press, 1991), 20; Clifford R. Kay, 'Oral Contraceptive Study of the Royal College of General Practitioners', in *Lecture Notes in Medical Informatics: Long-term Studies on the Side-Effects of Contraception—State and Planning, Symposium, Munich 1977*, eds. Ursula Kellhammer and Karl Überla (Berlin: Springer-Verlag, 1978), 29–30.
16. Hera Cook, *The Long Sexual Revolution: English Women, Sex, and Contraception, 1800-1975* (Oxford: Oxford University Press, 2004), 293.
17. K. Wellings, 'Trends in Contraceptive Usage Since 1970', *British Journal of Family Planning* 12, 2 (1986): 15–22.
18. M. Smith and P. Kane, *The Pill off Prescription* (London: Birth Control Trust, 1975), 7.
19. Borge, *Protective Practices*, 193–4.
20. Matt Cook, 'Sexual Revolution(s) in Britain', in *Sexual Revolutions*, eds. Gert Hekma and Alain Giami (Basingstoke: Palgrave Macmillan, 2014), 121.
21. Germaine Greer, *The Female Eunuch* (London: HarperCollins e-books, 2008), 315.
22. Kate Fisher, *Birth Control, Sex and Marriage in Britain, 1918–1960* (Oxford: Oxford University Press, 2006), 44–62; Cook, 'The English Sexual Revolution', 121.
23. Grant, *Sexing the Millennium*, 74.
24. Cook, 'The English Sexual Revolution', 121.
25. Andrew Rosen, *The Transformation of British Life, 1950–2000: A Social History* (Manchester: Manchester University Press, 2003), 52–6; Sarah Harper, 'The Challenge for Families of Demographic Ageing', in *Families in Ageing Societies: A Multi-Disciplinary Approach*, ed. Sarah Harper (Oxford: Oxford University Press, 2004), 8.
26. Kara Granzow, 'De-Constructing "Choice": The Social Imperative and Women's Use of the Birth Control Pill', *Culture, Health and Sexuality* 9, 1 (January–February 2007): 47–8.
27. Audrey Leathard, *Fight for Family Planning: The Development of Family Planning Services in Britain, 1921–74* (London: The Macmillan Press, 1980), 227.
28. James Merchant, ed., *Medical Views on Birth Control, with an Introduction by Sir Thomas Horder* (London: Martin Hopkins and Co. Ltd., 1926), xiii.
29. Leathard, *Fight for Family Planning*, 227.
30. Leathard, *Fight for Family Planning*, 227.
31. New Zealand Debates, *Parliamentary Debates: House of Representatives, Vol. 344* (Wellington: R. E. Owen Government Printer, 1966), 3173.
32. Cook, 'The English Sexual Revolution', 120.

Index

N. Szuhan, *The Family Planning Association and Contraceptive Science and Technology in Mid-Twentieth-Century Britain*, Medicine and Biomedical Sciences in Modern History, https://doi.org/10.1007/978-3-030-81300-0

FSC
www.fsc.org
MIX
Papier | Fördert gute Waldnutzung
FSC® C083411

Kolibri 360 GmbH
Ferdinand-Jühlke-Straße 7
99095 Erfurt, Deutschland
produktsicherheit@kolibri360.de